Topics in Autonomic Nervous System

Edited by María Elena Hernández-Aguilar and Gonzalo Emiliano Aranda-Abreu

Published in London, United Kingdom

Topics in Autonomic Nervous System
http://dx.doi.org/10.5772/intechopen.107768
Edited by María Elena Hernández-Aguilar and Gonzalo Emiliano Aranda-Abreu

Contributors
Sohel Mohamed Gamal Ahmed, Redha Waseem, Mogahed Ismail Hassan Hussein, Tayseer Salih Mohamed Salih, Dmitry Kruglov, Dermot McGuckin, Zeynep Balaban, Gökhan Kurt, Estela M. Muñoz, Martin Avila, Carlos L. Freites, Elena Vásquez, Juan B. Amiotti, Janina Borgonovo, Felipe Fanine de Souza, Felipe Ibiapina dos Reis, Julia Petry Trevisani, Mona Elsayed, Elizabeth Barbara Torres, Jorge Manzo, Deissy Herrera-Covarrubias, Flower M.J. Caycho Salazar, Genaro A. Coria-Ávila, Luis I. García-Hernández, María Rebeca Toledo-Cárdenas, María Elena Hernández Aguilar, Victoria Serhiyenko, Marta Hotsko, Yuriy Markevich, Martyn-Yurii Markevich, Volodymyr Segin, Ludmila Serhiyenko, Alexandr Serhiyenko

First published in London, United Kingdom, 2023 by IntechOpen
IntechOpen is the global imprint of INTECHOPEN LIMITED, registered in England and Wales, registration number: 11086078, 5 Princes Gate Court, London, SW7 2QJ, United Kingdom
Printed in Croatia

British Library Cataloguing-in-Publication Data
A catalogue record for this book is available from the British Library

Additional hard and PDF copies can be obtained from orders@intechopen.com

Topics in Autonomic Nervous System
Edited by María Elena Hernández-Aguilar and Gonzalo Emiliano Aranda-Abreu
p. cm.

This title is part of the Physiology Book Series, Volume 22
Topic: Human Physiology
Series Editor: Tomasz Brzozowski
Topic Editor: Kunihiro Sakuma
Associate Topic Editor: Kotomi Sakai

Print ISBN 978-1-83768-345-1
Online ISBN 978-1-83768-346-8
eBook (PDF) ISBN 978-1-83768-347-5
ISSN 2631-8261

For EU product safety concerns:
IN TECH d.o.o., Prolaz Marije Krucifikse Kozulić 3, 51000 Rijeka, Croatia, info@intechopen.com or visit our website at intechopen.com.

IntechOpen Book Series
Physiology
Volume 22

Aims and Scope of the Series

Modern physiology requires a comprehensive understanding of the integration of tissues and organs throughout the mammalian body, including the cooperation between structure and function at the cellular and molecular levels governed by gene and protein expression. While a daunting task, learning is facilitated by identifying common and effective signaling pathways mediated by a variety of factors employed by nature to preserve and sustain homeostatic life. As a leading example, the cellular interaction between intracellular concentration of Ca+2 increases, and changes in plasma membrane potential is integral for coordinating blood flow, governing the exocytosis of neurotransmitters, and modulating gene expression and cell effector secretory functions. Furthermore, in this manner, understanding the systemic interaction between the cardiovascular and nervous systems has become more important than ever as human populations' life prolongation, aging and mechanisms of cellular oxidative signaling are utilised for sustaining life. Altogether, physiological research enables our identification of distinct and precise points of transition from health to the development of multimorbidity throughout the inevitable aging disorders (e.g., diabetes, hypertension, chronic kidney disease, heart failure, peptic ulcer, inflammatory bowel disease, age-related macular degeneration, cancer). With consideration of all organ systems (e.g., brain, heart, lung, gut, skeletal and smooth muscle, liver, pancreas, kidney, eye) and the interactions thereof, this Physiology Series will address the goals of resolving (1) Aging physiology and chronic disease progression (2) Examination of key cellular pathways as they relate to calcium, oxidative stress, and electrical signaling, and (3) how changes in plasma membrane produced by lipid peroxidation products can affect aging physiology, covering new research in the area of cell, human, plant and animal physiology.

Meet the Series Editor

Prof. Dr. Thomas Brzozowski works as a professor of Human Physiology and is currently a Chairman at the Department of Physiology and is V-Dean of the Medical Faculty at Jagiellonian University Medical College, Cracow, Poland. His primary area of interest is physiology and pathophysiology of the gastrointestinal (GI) tract, with a major focus on the mechanism of GI mucosal defense, protection, and ulcer healing. He was a postdoctoral NIH fellow at the University of California and the Gastroenterology VA Medical Center, Irvine, Long Beach, CA, USA, and at the Gastroenterology Clinics Erlangen-Nuremberg and Munster in Germany. He has published 290 original articles in some of the most prestigious scientific journals and seven book chapters on the pathophysiology of the GI tract, gastroprotection, ulcer healing, drug therapy of peptic ulcers, hormonal regulation of the gut, and inflammatory bowel disease.

Meet the Volume Editors

Professor Dr. María Elena Hernández Aguilar works at the Brain Research Institute, Universidad Veracruzana, Xalapa, Veracruz, México. She received her BSc in Biology from Universidad Veracruzana, México, and her MSc and Ph.D. from the Universidad Nacional Autónoma de México in 1993 and 1997, respectively. Her research concentrates on molecular mechanisms involved in prostate cancer. She also investigates how the major pelvic ganglia may be involved in the generation of prostate cancer when neurons have degenerated. She coordinates Ph.D. theses in brain research studies and teaches neuroendocrinology to Ph.D. students. She is a member of the Southeastern Mexican chapter of the Society for Neuroscience and the Mexican Society of Urogenital Sciences. Dr. Aguilar is the recipient of the 2021 State Science and Technology Award.

Professor Dr. Gonzalo Emiliano Aranda Abreu works at the Brain Research Institute, Universidad Veracruzana, Xalapa, Veracruz, México. He received his BSc in Experimental Biology from the Universidad Autónoma Metropolitana, Iztalapa, México, in 1992. He obtained an MSc in Molecular Biology and Genetics from the Centro de Investigación y Estudios Avanzados, IPN, México, in 1996, and a Ph.D. in Neurobiology from Weizmann Institute of Science, Rehovot, Israel, in 2001. He is an expert in Alzheimer's disease research, principally analyzing tau protein transport in the neural axon and in relation to Alzheimer's disease. He has published book chapters on brain rehabilitation. He is president and member of the Southeastern Mexican chapter of the Society for Neuroscience. Dr. Abreu is involved in the training of doctoral students and teaches molecular and cellular neurobiology and bioinformatics to Ph.D. students.

Contents

Preface **XIII**

Section 1
Pharmacology and Neurotransmitters of the Autonomic Nervous System **1**

Chapter 1 **3**
Neurotransmitters of Autonomic Nervous System
by Zeynep Balaban and Gokhan Kurt

Chapter 2 **17**
Pharmacology of the Autonomic Nervous System
by Redha Waseem, Mogahed Ismail Hassan Hussein, Tayseer Salih Mohamed Salih and Sohel Mohamed Gamal Ahmed

Section 2
Autonomic Influences on Pain and Cardiovascular Responses **43**

Chapter 3 **45**
The Role of Autonomic Nervous System in Pain Chronicity
by Dmitry Kruglov and Dermot McGuckin

Chapter 4 **71**
Exploring Cardiac Responses of Pain and Distress
by Mona Elsayed and Elizabeth Barbara Torres

Section 3
Neural Control in Unique Systems **87**

Chapter 5 **89**
Sympathetic Innervation of the Mammalian Pineal Gland: Its Involvement in Ontogeny and Physiology, and in Pineal Dysfunction
by Martin Avila, Carlos L. Freites, Elena Vásquez, Juan B. Amiotti, Janina Borgonovo and Estela M. Muñoz

Chapter 6 103
The Brain-Like Enteric Nervous System
by Flower M.J. Caycho Salazar, Deissy Herrera-Covarrubias, Genaro A. Coria-Ávila, Luis I. García-Hernández, María Rebeca Toledo-Cárdenas, María Elena Hernández-Aguilar and Jorge Manzo

Section 4
Neurological Insights into Cardiovascular and Neuromuscular Conditions 119

Chapter 7 121
Diabetic Cardiac Autonomic Neuropathy: Link between Heart Rate Variability, Violated Blood Pressure Pattern, and Pulse Wave Velocity
by Victoria Serhiyenko, Marta Hotsko, Yuriy Markevich, Martyn-Yurii Markevich, Volodymyr Segin, Ludmila Serhiyenko and Alexandr Serhiyenko

Chapter 8 143
Clinical, Pathophysiological and Electrodiagnostic Aspects of Lambert-Eaton Myasthenic Syndrome
by Felipe Fanine de Souza, Julia Petry Trevisani and Felipe Ibiapina dos Reis

Preface

The autonomic nervous system is an extension of the central nervous system and connects the different organs of the body and regulates their functions. It is extremely complex, not only because of the nerves that emerge from the different segments of the spinal cord but also because of the plexuses and ganglia that connect. It is a difficult system to study due to the complexity of the nerves and the number of sympathetic, parasympathetic, or sensory fibers they contain. In addition, little is known about how peripheral diseases participate, the molecular events that are part of the molecular events used to induce peripheral diseases, and how these molecular events could alter the function of the hormonal system. Therefore, this book presents updated information on the structure, function, autonomic ganglia, and pathologies of the autonomic nervous system. This book is recommended for anyone who is studying this system and its functions.

The editors would like to express their gratitude to all the authors who contributed to this book, as well as to the staff at IntechOpen for the invitation to participate in this book project.

María Elena Hernández-Aguilar and Gonzalo Emiliano Aranda-Abreu
Brain Research Institute,
University of Veracruzana,
Xalapa, Veracruz, Mexico

Section 1

Pharmacology and Neurotransmitters of the Autonomic Nervous System

Chapter 1

Neurotransmitters of Autonomic Nervous System

Zeynep Balaban and Gokhan Kurt

Abstract

Autonomic nervous system (ANS) regulates the physiologic process in the body and has essential role in the systems such as blood pressure regulation, respiration, heart rate, and sexual arousal. ANS is divided into the sympathetic nervous system and the parasympathetic nervous system and regulates whole organism functions in the body. Although the main neurotransmitters in the ANS are norephinephrine, epinephrine, and acetilcholine, many other different agents and chemicals play an important role of the neurotransmitters function. These molecules act on many different receptors and sides. This chapter provides a detailed evaluation of neurotransmitters, related molecules, their receptors and how they function to maintain autonomic functions in both the central and peripheral parts of the systems.

Keywords: neurotransmitter, receptor, sympathetic, parasympathetic, physiology, cholinergic

1. Introduction

The autonomic nervous system (ANS) is a complex network of nerves responsible for regulating involuntary body functions such as heart rate, blood pressure, bladder function, digestion, and respiration. The ANS is divided into two main branches: the sympathetic nervous system and the parasympathetic nervous system, both of which use different neurotransmitters to carry out their functions. The sympathetic nervous system is responsible for initiating the "fight or flight" response by increasing heart rate and blood pressure, dilating the pupils, and redirecting blood flow from the digestive system to the muscles. This reaction is triggered in response to perceived threats and prepares the body for action. Conversely, the parasympathetic nervous system slows the heart rate and respiration, constricts the pupils, and increases blood flow to the digestive system. This results in a "rest and digest" response, that promotes relaxation and facilitates recovery from stress. The ANS is critical to maintaining homeostasis by balancing body functions and adapting to changes in the environment. It regulates several vital activities that are important for survival and helps us respond appropriately to different situations [1].

Neurotransmitters are the crucial mediators of interneuronal communication, responsible for transmitting signals between neurons and their target cells. On the website ANS, several neurotransmitters have been discovered, each with unique functions and roles, including acetylcholine, norepinephrine, dopamine, serotonin, and

 IntechOpen

neuropeptides [1]. In recent years, numerous neurotransmitters have been discovered to be involved in signal transduction, revealing the complicated and multifaceted nature of autonomic regulation. This chapter will review some of the recent discoveries in this field and highlight the functions and mechanisms of action of some important neurotransmitters.

Recent discoveries have shown that the ANS is organized in complex ways and that the function and structure of non-synaptic autonomic neuroeffectors is a crucial aspect of ANS regulation. In addition to classical neurotransmission, the concept of co-transmission has been introduced, in which multiple neurotransmitters can be released from a single neuron [2]. In addition, neuromodulation is another important concept in neuroscience that has contributed significantly to our understanding of ANS. Neuromodulators are chemicals that alter the activity of neurotransmitters and their receptors, thereby modulating the strength and efficacy of synaptic transmission [3]. Both co-transmission and neuromodulation are essential for the flexible and dynamic regulation of physiological processes and enable ANS to respond rapidly and appropriately to changes in the internal and external environment [2, 3]. Thanks to advances in molecular biology and imaging techniques, we can study these processes in great detail and gain insight into the complex mechanisms underlying autonomic neurotransmission.

2. Neuroeffector junction

The autonomic neuromuscular junction (ANMJ) is a specialized synapse where autonomic nerve impulses are transmitted to effector cells such as smooth muscle, cardiac muscle, and glands. Unlike the skeletal neuromuscular junction, the ANMJ lacks pre- and post-functional specialization. It has varicosities that release neurotransmitters during impulse transmission [4]. The structure of the ANMJ may vary depending on the type of effector cell. In smooth muscle cells, neuromuscular junctions are diffuse and distributed over a large area. In cardiac muscle cells, the neuromuscular junctions are located in the intercalated discs between adjacent cells. In glandular cells, the neuromuscular junction sites are located at the cell membrane. The ANMJ facilitates the transmission of nerve impulses through the release of neurotransmitters from the presynaptic neuron, the diffusion of these neurotransmitters across the synaptic cleft, and the activation of post-synaptic receptors on the effector cell. Acetylcholine and norepinephrine are the two major neurotransmitters involved in ANMJ [1, 5]. Acetylcholine is the neurotransmitter released by both preganglionic and post-ganglionic neurons of the parasympathetic division. It acts on muscarinic receptors in effector cells and causes smooth muscle cell contraction, slowing of heart rate in cardiac myocytes, and secretion of glandular cells [6]. In the sympathetic nervous system, norepinephrine is the neurotransmitter released by post-ganglionic neurons. Norepinephrine acts on alpha- and beta-adrenergic receptors in effector cells and leads to contraction of smooth muscle cells, acceleration of heart rate in cardiac muscle cells, and secretion of glandular cells [7].

The ANMJ effectors are muscle bundles connected by low-resistance pathways that allow electrotonic propagation of activity within the smooth muscle bundle. Varicosities are constantly in motion and have a dynamic relationship with muscle cell membranes, which means that a given impulse is likely to trigger transmitter release from only some of the varicosities it encounters. In addition, neurotransmitters such as dopamine, serotonin, and histamine may also be released at the ANMJ and modulate its activity. In addition, other substances such as hormones, locally released agents, and neurotransmitters from nearby nerves can also alter neurotransmission by affecting either the release or the action of the transmitter. Many of these substances, including co-transmitters, are capable of affecting neuronal growth and development. Because the autonomic neuroeffector junctions have a wide and variable gap, they are particularly suitable for the above mechanisms of neuronal control [8].

3. Signaling molecules and their receptors

Neurotransmitters are molecules released by nerves in response to electrical stimulation that bind to specific receptors on neighboring cells to produce a response. For a substance to be classified as a neurotransmitter, it must meet certain criteria, such as being synthesized and stored by the presynaptic neuron, being released in a calcium-dependent manner, having a mechanism to terminate release, and producing effects similar to those of electrical nerve stimulation when applied locally [9]. Although early studies identified only a few neurotransmitters in the ANS, more recent research has identified several substances, such as monoamines, amino acids, neuropeptides, ATP, nitric oxide (NO), and carbon monoxide (CO) [10, 11].

Neurons can store and release various neurotransmitters and neuromodulators that can have different effects on target cells. Neurons can have both slow-acting neuropeptide transmitters and fast-acting small molecule transmitters that can be present in the same neurons and released through co-localized synaptic vesicles, or they can be stored in different groups of vesicles to transmit signals together (**Table 1**) [12].

Research has also shown that neurotransmitters can have multiple functions within the ANS. Acetylcholine, for example, was previously thought to be exclusively responsible for parasympathetic signaling, whereas norepinephrine was thought to be responsible for sympathetic signaling. However, acetylcholine can also be released from sympathetic neurons and act as a modulator of sympathetic activity [6].

ANS has two types of receptors: cholinergic and adrenergic. Acetylcholine activates the cholinergic receptors, while catecholamines such as epinephrine and norepinephrine activate the adrenergic receptors. Cholinergic receptors are divided into two categories: nicotinic receptors and muscarinic receptors. Nicotinic receptors are mainly located in the autonomic ganglia and neuromuscular junction, whereas muscarinic receptors are located in the effector organs of the PNS and some tissues innervated by the SNS. Adrenergic receptors are divided into alpha and beta receptors. Alpha receptors have two subtypes, alpha-1 and alpha-2 receptors. Alpha-1 receptors are located in the smooth muscle of blood vessels and in the iris of the eye, while alpha-2 receptors are located in presynaptic neurons and inhibit the release of norepinephrine. Beta receptors are also of two subtypes, beta-1 and beta-2 receptors. Beta-1 receptors are located in the heart, while beta-2 receptors are found in lung smooth muscle and skeletal muscle. ANS has regulatory systems such as self-inhibition of norepinephrine release via presynaptic alpha-2 receptors, regulation of norepinephrine synthesis, and desensitization and hypersensitization of adrenoceptors. Acetylcholine acts on two classes of receptors: nicotinic receptors, found mainly in ganglia, and muscarinic receptors, which are coupled to G proteins and respond more slowly. Among purine receptors, there are two main types: P1 receptors, which are sensitive to adenosine and blocked by methylxanthines, and P2 receptors, which are sensitive to ATP and can lead to prostaglandin synthesis.

Neurotransmitter	Category	Function
Acetylcholine (ACh)		Contraction of smooth muscle cells, slowing of heart rate in cardiac muscle cells, secretion of glandular cells
Norepinephrine (NE)	Catecholamines	Contraction of smooth muscle cells, acceleration of heart rate in cardiac muscle cells
Serotonin	Indolamines	Modulation of autonomic function
Gamma-aminobutyric acid (GABA)	Amino acids	Inhibition of autonomic function
Glutamate	Amino acids	Excitation of autonomic function
Adenosine triphosphate (ATP)	Purines	Modulation of autonomic function
Nitric oxide (NO)	Soluble gases	Modulation of autonomic function

Some neurotransmitters can be both excitatory and inhibitory depending on the receptor to which they bind, and their functions can vary depending on the specific location and target of the autonomic nervous system.

Table 1.
Neurotransmitters of the Autonomic Nervous System.

Neuropeptide receptors are G protein-coupled receptors that activate adenylyl cyclase or phospholipase C as signal transducers [13–15].

3.1 Acetylcholine

Acetylcholine is a chemical messenger produced by neurons in various parts of the nervous system. It is released by large pyramidal cells in the motor cortex, various neurons in the basal ganglia, and motor neurons controlling skeletal muscles, among others. Acetylcholine usually has a stimulatory effect on nerve cells, but it can also inhibit certain peripheral parasympathetic nerves, such as those that slow the heart. Choline acetyltransferase (ChAT) is the enzyme responsible for the synthesis of acetylcholine from choline and acetyl coenzyme A in the cytoplasm of nerve cells. After production, acetylcholine is stored in tiny vesicles that have a specific transporter in their membrane. When electrical signals trigger the release of calcium ions, acetylcholine is released into the synaptic cleft, where it can bind to receptors on nearby cells. Acetylcholinesterase is an enzyme that breaks down acetylcholine, limiting its action. The breakdown of acetylcholine produces choline, which is then transported back into neurons to form more acetylcholine. The uptake of choline into the presynaptic terminal is a crucial step in the production of acetylcholine [6].

3.2 Norepinephrine

The neurotransmitter norepinephrine is synthesized by three enzymes and released by neurons in the brainstem and hypothalamus, particularly in the locus ceruleus of the pons. Terminals of these neurons release norepinephrine into the extracellular space by exocytosis triggered by electrical stimulation and a Ca^{2+}-dependent process. Norepinephrine is stored in small and large dense nuclear vesicles in the neuronal cytosol alongside chromogranins and dopamine-β-hydroxylase. Its action is rapidly terminated when it interacts with specific receptors by being recycled into neuronal nodes or non-neuronal cells. Monoamine oxidase (MAO) and catechol-O-methyltransferase (COMT) metabolize norepinephrine in intracellular cells. Norepinephrine controls overall activity and mood of the mind by transmitting nerve fibers to different brain areas, resulting in increased alertness. In most of these areas, norepinephrine activates excitatory receptors, although in some regions it triggers inhibitory receptors instead. The post-ganglionic neurons of the SNS secrete NE, which has both excitatory and inhibitory effects on various organs [1, 16].

3.3 ATP

Adenosine triphosphate (ATP), a type of purine nucleotide, was originally identified as the first neurotransmitter in nonadrenergic noncholinergic (NANC) nerves that met the criteria for a neurotransmitter. Further research has shown that purinergic signaling is widespread in both neural and non-neural systems. ATP acts as a neurotransmitter at neuroeffectors, synapses in peripheral autonomic ganglia, and in the brain and spinal cord. It also plays a critical role as a signaling molecule in the enteric nervous system and on sensory nerves affecting both physiological reflexes and nociception. ATP is synthesized in nerve terminals and stored in vesicles. Once released, it binds to post-functional P2X ion channel receptors and is rapidly cleaved by ectonucleotidases into adenosine diphosphate (ADP), adenosine monophosphate (AMP), and adenosine. Adenosine is then transported back into neurons and non-neuronal cells via a high-affinity nucleoside carrier uptake system. It can be converted back to ATP and returned to vesicles or further degraded by adenosine deaminase to inosine, which is inactive and enters the bloodstream. Adenosine acts on prefunctional P1 receptors and inhibits neurotransmitter release. In addition, small amounts of other nucleotides such as ADP, AMP, guanosine triphosphate (GTP), uridine triphosphate (UTP) and diadenosine polyphosphates have been detected in synaptic vesicles, which may also have a neuromodulatory function in the nervous system [3, 17].

3.4 Nitric oxide

NO is a putative neurotransmitter in the ANS that is synthesized in a reaction in which L-arginine is converted to L-citrulline by nitric oxide synthase (NOS). Unlike other neurotransmitters, NO is not stored in vesicles but is synthesized almost instantaneously on demand and diffuses from presynaptic terminals to act on intracellular guanylate cyclase in the post-synaptic neuron, resulting in relaxation. Type I NOS, which is constitutively expressed in autonomic neurons, is stimulated by Ca^{2+} during transmission. NO does not act on extracellular receptors but rather at intracellular sites, and its unstable nature allows it to terminate NO-dependent responses without the need for degradative enzymes or reuptake. In addition, NO can readily bind to the heme group of hemoglobin and inhibits NO-dependent reactions. Nitric oxide is not only found in the autonomic nervous system, but is also produced by nerve terminals in brain regions responsible for long-term behavior and memory. Its unique mechanism of formation in the presynaptic terminal and its action on the post-synaptic neuron distinguish it from other small molecule transmitters. NO is synthesized almost immediately and diffuses out of the presynaptic terminals within seconds rather than being released in vesicular packets. Once in the post-synaptic neuron, it does not significantly alter membrane potential but modifies intracellular metabolic functions to alter neuronal excitability for seconds, minutes, or possibly even longer. Therefore, NO could shed light on previously unexplained behavioral and memory functions [18, 19].

3.5 Other neurotransmitters

The ANS uses several neurotransmitters to regulate various physiological functions. One of these neurotransmitters is 5-hydroxytryptamine (5-HT), which is synthesized from tryptophan via 5-hydroxytryptophan by tryptophan hydroxylase and l-aromatic amino acid decarboxylase. Hydroxytryptamin (HT) While 5-hydroxytryptophan is primarily synthesized in myenteric neurons, it can also act as a spurious neurotransmitter after being taken up and released by sympathetic nerves. Similarly, dopamine, GABA, and glutamate, which are classic neurotransmitters in the central nervous system, also act as autonomic neurotransmitters. GABA is the major inhibitory neurotransmitter in the adult central nervous system and is secreted by nerve terminals in the spinal cord, cerebellum, basal ganglia, and many areas of the cortex. The role of GABA in enteric neurotransmission has been identified, where it acts through excitatory GABAA and prefunctional inhibitory GABAB receptors. Dopamine is secreted by neurons from the substantia nigra, terminating mainly in the striatal region of the basal ganglia. Its action is primarily inhibitory. Glutamate is secreted from presynaptic terminals in many sensory pathways entering the central nervous system and in many areas of the cerebral cortex and is known to cause excitation. Serotonin, secreted by nuclei in the median raphe of the brainstem, acts as an inhibitor of pain pathways in the spinal cord and contributes to mood control and sleep initiation in higher regions of the nervous system [1, 20].

3.5.1 Hydrogen sulfide (H2S)

H2S is a colorless, flammable gas that has long been known as a toxic environmental pollutant. However, recent studies have shown that H2S is also an endogenously produced gasotransmitter that plays a crucial role in various physiological processes in the body. H2S is produced by the enzyme cystathionine beta synthase as part of the transsulfuration pathway that converts homocysteine to cysteine. Another enzyme, cystathionine gamma lyase, can also produce H2S from cysteine. The third H2S-producing enzyme, 3-mercaptopyruvate sulfurtransferase, produces H2S from 3-mercaptopyruvate. H2S can also be produced by the gut microbiota, which metabolizes sulfur-containing amino acids. Once produced, H2S acts as a signaling molecule that regulates various physiological processes, including blood pressure, inflammation, and cell signaling. H2S has also been shown to have anti-inflammatory, antioxidant, and cytoprotective effects. One of the most important physiological processes regulated by H2S is vasodilation, which contributes to the regulation of blood pressure. H2S induces vasodilation by activating ATP-sensitive potassium channels in vascular smooth muscle cells. This leads to hyperpolarization of the cell membrane, resulting in smooth muscle cell relaxation and subsequent vasodilation. H2S also has an anti-inflammatory effect. It can inhibit the production of pro-inflammatory cytokines such as interleukin-1 beta (IL-1B), tumor necrosis factor-alpha (TNF-alpha), and interleukin-6 (IL-6). This anti-inflammatory effect is thought to be mediated by inhibiting the activation of nuclear factor kappa B (NF-kB), which is an important regulator of the inflammatory response. In addition, H2S has been shown to have a cytoprotective effect. It can protect cells from oxidative stress-induced damage and apoptosis. H2S can also increase the activity of antioxidant enzymes such as superoxide dismutase and catalase, which also contributes to its cytoprotective effect. Overall, H2S is a gasotransmitter that plays a crucial role in regulating various physiological processes in the body. Its vasodilatory, anti-inflammatory, and cytoprotective effects make it a promising therapeutic target for the treatment of various diseases such as hypertension, inflammation, and oxidative stress-related disorders [21, 22].

3.5.2 Neuropeptide Y

NPY is a 36 amino acid neuropeptide widely distributed in the central and peripheral nervous system. It belongs to the peptide family, which also includes peptide YY (PYY) and pancreatic polypeptide (PP). NPY acts as a neurotransmitter in the brain, where it is involved in a number of physiological functions, including appetite regulation, stress response, anxiety, and mood regulation. NPY is also found in the peripheral nervous system, where it regulates cardiovascular function, gastrointestinal motility, and immune function. In humans, the NPY gene is located on chromosome 7, and the peptide is synthesized in the cell bodies of neurons in the brain and peripheral nervous system. NPY is released by nerve terminals in response to a variety of stimuli, including stress, fasting, and exercise. NPY exerts its effects by binding to a family of G protein-coupled receptors called Y receptors. There are five known Y receptors (Y1, Y2, Y4, Y5, and Y6), each of which has a different distribution pattern in the brain and peripheral tissues. The Y1 receptor is the most abundant subtype in the brain and is involved in the regulation of feeding behavior, anxiety, and pain perception. The Y2 receptor is also found in the brain and is involved in modulating the release of neurotransmitters. The Y4 and Y5 receptors are mainly found in the periphery, where they regulate food intake, adiposity, and glucose homeostasis. The Y6 receptor is expressed in the brain, but its function is not well understood. NPY is associated with a number of human diseases, including obesity, diabetes, anxiety, and cardiovascular disease. In obesity, elevated levels of NPY have been observed, leading to increased food intake and weight gain. In diabetes, NPY has been shown to play a role in regulating glucose homeostasis, and drugs targeting the Y2 receptor have been suggested as potential therapies [23–25].

3.5.3 Orexin

Another recently discovered ANS neurotransmitter is orexin, also known as hypocretin. It is a neuropeptide produced mainly in a small group of neurons in the hypothalamus of the brain. This neuropeptide plays an important role in regulating various physiological processes, including sleep, wakefulness, feeding behavior, energy homeostasis, and reward systems.

Orexin was first discovered in 1998, and since then, extensive research has been conducted to understand its functions in the brain. One of the most important roles of orexin is its involvement in sleep regulation. Orexin neurons are active during periods of wakefulness and promote wakefulness by stimulating the release of other neurotransmitters such as dopamine, norepinephrine, and histamine. These neurotransmitters ensure that the brain remains in a state of arousal and alertness. In addition, orexin has been found to play a critical role in regulating feeding behavior and energy homeostasis. Orexin promotes feeding behavior by increasing appetite and enhancing the rewarding properties of food. This neuropeptide also regulates energy expenditure by increasing thermogenesis, or heat production, in brown adipose tissue. Studies also suggest that orexin may be involved in the development of addiction and drug-seeking behavior. This neuropeptide has been found to enhance the rewarding effects of drugs such as cocaine and amphetamines by stimulating the release of dopamine in the brain's reward centers. In addition, recent studies have shown that orexin may play a role in regulating emotional behaviors such as anxiety and depression. Orexin signaling has been found to be disrupted in people with anxiety and depressive disorders, and modulation of orexin signaling has been suggested as a potential therapeutic target for these disorders [26, 27].

3.5.4 PACAP

Pituitary adenylate-cyclase-activating polypeptide (PACAP) is a neuropeptide that acts as a neurotransmitter or neuromodulator in the central and peripheral nervous systems. It was first discovered in the hypothalamus, where it has been shown to stimulate the release of adrenocorticotropic hormone from the pituitary gland. Since then, PACAP has been found to have a variety of functions in the nervous system, including regulating the release of neurotransmitters, modulating synaptic plasticity, and maintaining neuronal survival. One of the most important functions of PACAP in the nervous system is the regulation of neurotransmitter release. PACAP has been shown to stimulate the release of several neurotransmitters, including acetylcholine, norepinephrine, and dopamine, from both central and peripheral neurons. This suggests that PACAP may play an important role in regulating autonomic function as well as modulating higher brain functions such as learning and memory. PACAP also plays a critical role in synaptic plasticity, the process by which the strength of synapses between neurons is altered in response to changes in neuronal activity. In particular, PACAP has been shown to enhance long-term potentiation, a form of synaptic plasticity thought to underlie learning and memory. This suggests that PACAP could be an important target for the development of drugs to improve cognitive function. Another important function of PACAP in the nervous system is to maintain neuron survival. PACAP has been shown to protect neurons from a range of damage, including oxidative stress, ischemia, and excitotoxicity [28, 29]. This suggests that PACAP may have therapeutic potential for the treatment of neurodegenerative diseases such as Alzheimer's and Parkinson's disease.

3.5.5 Galanin

Recent research has identified the neuropeptide galanin as another important ANS neurotransmitter. Galanin is a neuropeptide widely distributed in the central and peripheral nervous system. It was first discovered in the pig intestine in 1983, but later studies showed that it is also expressed in the brain and various other tissues. Galanin is synthesized as a precursor protein and then cleaved into smaller peptides that are released by nerve terminals as neurotransmitters or neuromodulators. Galanin acts on three different G protein-coupled receptors, namely GAL1, GAL2, and GAL3. These receptors are widely distributed in the brain and peripheral tissues, suggesting that galanin has multiple biological effects. In the nervous system, galanin is involved in the regulation of a variety of functions, including feeding, pain perception, memory, and anxiety. One of the most important functions of galanin in the nervous system is its involvement in pain modulation. Galanin has been shown to inhibit the release of substance P, a neuropeptide involved in the transmission of pain signals. Galanin also regulates the activity of nociceptors, the primary sensory neurons that respond to painful stimuli. These effects of galanin suggest that it is a potential therapeutic agent for the treatment of chronic pain. In addition to its role in pain modulation, galanin is also involved in the regulation of feeding behavior. Studies have shown that galanin stimulates feeding behavior in animals, and blocking its activity can lead to decreased food intake and weight loss. The GAL1 receptor has been identified as the main mediator of galanin's effects on feeding behavior, making it a potential target for the treatment of obesity. Galanin has also been associated with the regulation of memory and anxiety. Studies have shown that galanin levels are altered in the brains of animals exposed to stress and that administration of galanin can attenuate the behavioral and neurochemical effects of stress. These results suggest that galanin may play a protective role against the negative effects of stress on the brain and that it has potential for treating anxiety [30–32].

3.5.6 Taurine

Taurine, an amino acid with neuroprotective properties, plays a crucial role in regulating various cellular processes in the central nervous system. Taurine acts as a neuromodulator within the ANS, affecting neuronal excitability and autonomic functions. It has been shown to have a significant impact on neural stem and progenitor cells through modulation of gene expression. Taurine exerts its protective effects by influencing inflammatory processes in the central nervous system, inhibiting apoptosis, acting as an antioxidant, and controlling cell volume and water content in neurons. One of the most important mechanisms by which taurine provides neuroprotection is the suppression of apoptosis or programmed cell death. Taurine acts on both ionotropic taurine receptors and metabotropic taurine receptors to inhibit apoptosis triggered by stress in the endoplasmic reticulum (ER). By attenuating apoptosis triggered by ER, taurine helps to ensure neuron survival and prevent neuronal damage. In addition, taurine has antioxidant properties that effectively scavenge free radicals and reduce oxidative stress in the central nervous system. This antioxidant activity helps protect neurons from oxidative damage and contributes to the overall neuroprotective effects of taurine. The neuroprotective properties of taurine have made it a promising candidate for the prophylaxis and treatment of neurodegenerative diseases [33–37].

4. Conclusion

The ANS plays a critical role in maintaining homeostasis and regulating physiological functions throughout the body. Neurotransmitters serve as important mediators in the transmission of signals within the ANS and enable precise communication between neurons and their target tissues or organs. In this comprehensive exploration of the neurotransmitters of the autonomic nervous system, we have gained valuable insights into their intricate mechanisms and physiological effects. Recent advances in our understanding of the ANS have been driven by the identification of new neurotransmitters and their functions. These findings have led to a more comprehensive understanding of the complex mechanisms that regulate various physiological processes. The discovery of new ANS neurotransmitters has opened new possibilities for targeted treatments of autonomic disorders. Precise targeting of these neurotransmitters could lead to more effective therapies with fewer side effects than current treatments. In addition, the discovery of new neurotransmitters has shed light on the intricate signal transduction pathways underlying ANS regulation. By studying these pathways, researchers can gain a deeper understanding of how different physiological systems interact to maintain homeostasis in the body. As research in this area continues to advance, we can expect to gain further insight into the functions of ANS neurotransmitters and their role in regulating physiological processes. These discoveries promise new treatments and therapies for autonomic disorders, as well as a more comprehensive understanding of the complexity of the autonomic nervous system.

Author details

Zeynep Balaban* and Gokhan Kurt
Department of Neurosurgery, Faculty of Medicine, Gazi University, Ankara, Turkey

*Address all correspondence to: zeynep.balaban@yahoo.com

References

[1] Guyton AC, Hall JE. Textbook of Medical Physiology. Philadelphia: WB Saunders; 2016. pp. 741-747

[2] Svensson E, Apergis-Schoute J, Burnstock G, Nusbaum MP, Parker D, Schiöth HB. General principles of neuronal co-transmission: Insights from multiple model systems. Frontiers in Neural Circuits. 2019;**12**:117. DOI: 10.3389/fncir.2018.00117

[3] Lundberg JM. Pharmacology of cotransmission in the autonomic nervous system: Integrative aspects on amines, neuropeptides, adenosine triphosphate, amino acids and nitric oxide. Pharmacological Reviews. 1996;**48**(1):113-178

[4] Hillarp NÅ. Structure of the synapse and the peripheral innervation apparatus of the autonomic nervous system. Acta Anatomica. 1946;**4**:1-153

[5] Jones RA, Harrison C, Eaton SL, Llavero Hurtado M, Graham LC, Alkhammash L, et al. Cellular and molecular anatomy of the human neuromuscular junction. Cell Reports. 2017;**21**(9):2348-2356. DOI: 10.1016/j.celrep.2017.11.008

[6] Sam C, Bordoni B. Physiology, acetylcholine. April 14, 2022. In: StatPearls [Internet]. Treasure Island (FL): StatPearls Publishing; 2023

[7] Hussain LS, Reddy V, Maani CV. Physiology, noradrenergic synapse. May 8, 2022. In: StatPearls [Internet]. Treasure Island (FL): StatPearls Publishing; 2023

[8] Burnstock G. Autonomic neuromuscular junctions: Current developments and future directions. Journal of Anatomy. 1986;**146**:1-30

[9] Inoue K. Neurotransmitter. In: Binder MD, Hirokawa N, Windhorst U, editors. Encyclopedia of Neuroscience. Berlin, Heidelberg: Springer; 2009. DOI: 10.1007/978-3-540-29678-2_3952

[10] Kennedy C. ATP as a cotransmitter in the autonomic nervous system. Autonomic Neuroscience. 2015;**191**:2-15. DOI: 10.1016/j.autneu.2015.04.004

[11] Xue L, Farrugia G, Miller SM, Ferris CD, Snyder SH, Szurszewski JH. Carbon monoxide and nitric oxide as coneurotransmitters in the enteric nervous system: Evidence from genomic deletion of biosynthetic enzymes. Proceedings of the National Academy of Sciences of the United States of America. 2000;**97**(4):1851-1855. DOI: 10.1073/pnas.97.4.1851

[12] Hnasko TS, Edwards RH. Neurotransmitter corelease: Mechanism and physiological role. Annual Review of Physiology. 2012;**74**:225-243. DOI: 10.1146/annurev-physiol-020911-153315

[13] McCorry LK. Physiology of the autonomic nervous system. American Journal of Pharmaceutical Education. 2007;**71**(4):78. DOI: 10.5688/aj710478

[14] Burnstock G. A basis for distinguishing two types of purinergic receptor. In: Straub RW, Bolis L, editors. Cell Membrane Receptors for Drugs and Hormones: A Multidisciplinary Approach. New York: Raven Press; 1978. pp. 107-118

[15] Brain SD, Cox HM. Neuropeptides and their receptors: Innovative science providing novel therapeutic targets. British Journal of Pharmacology. 2006;**147**(Suppl 1):S202-S211. DOI: 10.1038/sj.bjp.0706461

[16] Fillenz M. Transmission: Noradrenaline. Autonomic Neuroeffector Mechanisms. 1995;**4**:323-365

[17] Burnstock G. Historical review: ATP as a neurotransmitter. Trends in Pharmacological Sciences. 2006;**27**(3):166-176. DOI: 10.1016/j.tips.2006.01.005

[18] Kuriyama K, Ohkuma S. Role of nitric oxide in central synaptic transmission: Effects on neurotransmitter release. Japanese Journal of Pharmacology. 1995;**69**(1):1-8. DOI: 10.1254/jjp.69.1

[19] Togo T, Katsuse O, Iseki E. Nitric oxide pathways in Alzheimer's disease and other neurodegenerative dementias. Neurological Research. 2004;**26**(5):563-566. DOI: 10.1179/016164104225016236

[20] Burnstock G. Autonomic neurotransmission: 60 years since sir Henry dale. Annual Review of Pharmacology and Toxicology. 2009;**49**:1-30. DOI: 10.1146/annurev.pharmtox.052808.102215

[21] Kimura H. Hydrogen sulfide as a neuromodulator. Molecular Neurobiology. 2002;**26**(1):13-19. DOI: 10.1385/MN:26:1:013

[22] Hosoki R, Matsuki N, Kimura H. The possible role of hydrogen sulfide as an endogenous smooth muscle relaxant in synergy with nitric oxide. Biochemical and Biophysical Research Communications. 1997;**237**(3):527-531. DOI: 10.1006/bbrc.1997.6878

[23] Beck B. Neuropeptide Y in normal eating and in genetic and dietary-induced obesity. Philosophical Transactions of the Royal Society of London. Series B, Biological Sciences. 2006;**361**(1471):1159-1185. DOI: 10.1098/rstb.2006.1855

[24] Gehlert DR. Introduction to the reviews on neuropeptide Y. Neuropeptides. 2004;**38**(4):135-140. DOI: 10.1016/j.npep.2004.07.002

[25] Yi M, Li H, Wu Z, Yan J, Liu Q, Ou C, et al. A promising therapeutic target for metabolic diseases: Neuropeptide Y receptors in humans. Cellular Physiology and Biochemistry. 2018;**45**(1):88-107. DOI: 10.1159/000486225

[26] Wang C, Wang Q, Ji B, Pan Y, Xu C, Cheng B, et al. The Orexin/receptor system: Molecular mechanism and therapeutic potential for neurological diseases. Frontiers in Molecular Neuroscience. 2018;**11**:220. DOI: 10.3389/fnmol.2018.00220

[27] Martynska L, Wolinska-Witort E, Chmielowska M, Bik W, Baranowska B. The physiological role of orexins. Neuro Endocrinology Letters. 2005;**26**(4):289-292

[28] Hirabayashi T, Nakamachi T, Shioda S. Discovery of PACAP and its receptors in the brain. The Journal of Headache and Pain. 2018;**19**(1):28. DOI: 10.1186/s10194-018-0855-1

[29] Arimura A. Perspectives on pituitary adenylate cyclase activating polypeptide (PACAP) in the neuroendocrine, endocrine, and nervous systems. The Japanese Journal of Physiology. 1998;**48**(5):301-331. DOI: 10.2170/jjphysiol.48.301

[30] Lundström L, Elmquist A, Bartfai T, Langel U. Galanin and its receptors in neurological disorders. Neuromolecular Medicine. 2005;**7**(1-2):157-180. DOI: 10.1385/NMM:7:1-2:157

[31] Baranowska B et al. Neuropeptide Y, galanin, and leptin release in obese women and in women

with anorexia nervosa. Metabolism. 1997;**46**(12):1384-1389

[32] Lang R, Gundlach AL, Kofler B. The galanin peptide family: Receptor pharmacology, pleiotropic biological actions, and implications in health and disease. Pharmacology & Therapeutics. 2007;**115**(2):177-207. DOI: 10.1016/j.pharmthera.2007.05.009

[33] Menzie J, Prentice H, Wu J-Y. Neuroprotective mechanisms of taurine against ischemic stroke. Brain Sciences. 2013;*3*:877-907. DOI: 10.3390/brainsci3020877

[34] Paula-Lima AC et al. Activation of GABAA receptors by taurine and muscimol blocks the neurotoxicity of β-amyloid in rat hippocampal and cortical neurons. Neuropharmacology. 2005;**49**(8):1140-1148

[35] Schaffer S, Kim HW. Effects and mechanisms of Taurine as a therapeutic agent. Biomolecules & Therapeutics. 2018;**26**(3):225-241

[36] El Idrissi A, Trenkner E. Taurine as a modulator of excitatory and inhibitory neurotransmission. Neurochemical Research. 2004;**29**(1):189-197

[37] Wu HJ, Wu D, Xu Y. Effects of taurine on neuron protection. World Journal of Biological Chemistry. 2015;**6**(2):57-64

Chapter 2

Pharmacology of the Autonomic Nervous System

Redha Waseem, Mogahed Ismail Hassan Hussein, Tayseer Salih Mohamed Salih and Sohel Mohamed Gamal Ahmed

Abstract

This comprehensive chapter delves into the intricate landscape of autonomic nervous system (ANS) pharmacology. It meticulously explores both agonists and antagonists pharmacology that work on the sympathetic and parasympathetic divisions. This chapter covers direct and indirectly acting drugs and thoroughly explains receptor interactions. The content spans a wide array of examples, elucidating these agents' mechanisms and clinical applications. Through a detailed examination of pharmacokinetics, metabolism, adverse effects, and contraindications, healthcare professionals gain the insights needed to navigate the complexities of ANS modulation. This knowledge equips practitioners to harness the potential of autonomic drugs, facilitating optimal therapeutic outcomes across diverse medical scenarios.

Keywords: pharmacology, autonomic nervous system, sympathetic, parasympathetic, medications

1. Introduction

The autonomic nervous system (ANS) plays a pivotal role in maintaining homeostasis by modulating various vital processes, including heart rate, blood pressure, respiratory rate, gastrointestinal motility, and glandular secretions [1]. Understanding the pharmacology of the ANS is paramount in medicine, particularly in anesthesia and other acute medical specialities, as it allows healthcare professionals to manipulate autonomic pathways effectively and achieve desirable clinical outcomes [2]. This chapter aims to provide a comprehensive overview of ANS pharmacology, focusing on the sympathetic (SANS) and parasympathetic (PANS) divisions and their associated receptors.

Understanding the receptor selectivity of pharmacological agents is paramount in achieving desired clinical outcomes. Many drugs exhibit selectivity for specific adrenergic or cholinergic receptors, allowing for targeted modulation of the SANS and PANS [3]. Healthcare professionals can manipulate autonomic pathways to optimize patient care by carefully selecting and administering peripheral nervous system agonists or antagonists.

The knowledge of ANS pharmacology is particularly crucial in acute medical specialities, where precise control over the cardiovascular system, airway dynamics, and other physiological parameters is essential. Physicians rely on drugs that selectively target specific adrenergic or cholinergic receptors to achieve optimal hemodynamic stability and other vital parameters [2].

Furthermore, pharmacists, physicians, intensivists, and medical students benefit from a comprehensive understanding of ANS pharmacology. By grasping the complexities of autonomic receptor modulation, healthcare professionals can make informed decisions regarding drug selection, dosing, and potential adverse effects. This knowledge enhances patient safety and improves clinical outcomes across various medical disciplines.

2. Pharmacology of the sympathetic nervous system

The SANS, often referred to as the "fight or flight" system, prepares the body for physical exertion and stressful situations. The primary neurotransmitter in the SANS is norepinephrine (noradrenaline), which interacts with adrenergic receptors located throughout the body. These receptors are categorized into two main subtypes: α and β [1].

The α-adrenergic receptors are further divided into α-1 and α-2 subtypes. α-1 receptors are predominantly located in blood vessels, leading to vasoconstriction when activated. This effect increases systemic vascular resistance, elevating blood pressure. α-1 agonists such as phenylephrine find clinical utility in managing hypotension during anesthesia. Conversely, α-1 antagonists like prazosin induce vasodilation and alleviate conditions such as benign prostatic hyperplasia and hypertension. α-2 receptors are primarily located presynaptically in sympathetic nerve terminals, where their activation inhibits the release of norepinephrine, resulting in negative feedback regulation of sympathetic outflow. Clonidine, an α-2 agonist, is commonly used in anesthesia and surgery to attenuate sympathetic responses, promote sedation, and enhance perioperative analgesia [1].

β-adrenergic receptors consist of three subtypes: β-1, β-2, and β-3. β-1 receptors are predominantly found in the heart, activating heart rate and contractility [1]. β-1 agonists like dobutamine enhance cardiac output in patients with heart failure or cardiogenic shock. β-1 antagonists, such as metoprolol, are widely used in to mitigate the adverse effects of excessive sympathetic stimulation on the cardiovascular system [3]. β-2 receptors are abundant in the bronchial smooth muscle and peripheral vasculature, leading to bronchodilation and vasodilation when stimulated [1]. β-2 agonists like salbutamol are commonly utilized to manage asthma and chronic obstructive pulmonary disease (COPD) [3].

Conversely, β-2 antagonists may be used in conditions like glaucoma, where reduced intraocular pressure is desirable [3]. β-3 receptors are primarily present in adipose tissue, where their activation promotes lipolysis. While the therapeutic significance of β-3 receptors is still being explored, their modulation may hold potential in treating obesity and metabolic disorders.

2.1 Sympathomimetics

2.1.1 α-1 receptor agonists

α-1 receptors dominate most of the smooth muscle of the autonomic target organs. They mediate primarily arterial and venous vasoconstriction when activated. Drugs that mimic the action of epinephrine and norepinephrine can be called sympathomimetics. These drugs can be divided into direct and indirect agonists [4].

Direct agonists interact with the adrenoreceptor directly and subsequently activate them, while indirect agonists depend on their ability to enhance the effect of endogenous catecholamines. The indirect agonist can do so by (i) displacing catecholamine from their adrenergic nerve endings and inducing their release (e.g., the mechanism of action of tyramine), (ii) inhibiting the clearance of catecholamines by decreasing their neuronal reuptake (e.g., the mechanism of action of cocaine and tricyclic antidepressants), or (iii) preventing the enzymatic metabolism of norepinephrine (monoamine oxidase and catechol-O-methyltransferase inhibitors) [4].

2.1.1.1 Direct-acting α-agonists

2.1.1.1.1 Phenylephrine

The chemical formula of phenylephrine is $C_9H_{13}NO_2$. It is an α-1 adrenergic agonist that only affects β receptors at very high doses. As it is not a catechol derivative, it is not broken down by catechol-O-methyltransferase (COMT) and has a longer duration of action than catecholamines. It can cause an increase in blood pressure by venous and arteriolar vasoconstriction, and since it does not act on β receptors, there is no direct effect on cardiac muscle. The increase in blood pressure causes reflex bradycardia by stimulation of baroreceptors [5].

The intravenous (IV) phenylephrine hydrochloride increases blood pressure in adults with clinically significant hypotension resulting primarily from vasodilation in such settings as septic shock or anesthesia. Phenylephrine hydrochloride (HCL) is also used over-the-counter in ophthalmic formulations to promote mydriasis and conjunctival blood vessel vasoconstriction, intranasal administration as a treatment for uncomplicated nasal congestion, and as an over-the-counter additive to topical hemorrhoid medications [5].

The ophthalmic formulations of phenylephrine act for 3–8 hours, while intravenous solutions have a practical half-life of 5 minutes and an elimination half-life of 2.5 hours. The bioavailability orally is 38%, and ophthalmic solutions have clinically significant absorption, especially if the cornea is damaged. This drug is mainly metabolized by monoamine oxidase A, monoamine oxidase B, and sulfotransferase family 1A member 3 (SULT1A3). The primary metabolite it forms is the inactive meta-hydroxymandelic acid, followed by sulfate conjugates. It can also be metabolized to phenylephrine glucuronide. About 86% of the drug is recovered in urine; 16% of it is unmetabolized, and 57% of it is inactive meta-hydroxymendelic acid, and 8% is inactive sulfate conjugates [6].

The adverse effects of these drugs are nausea, vomiting, and confusion. Since phenylepinephrine increases the afterload more than the preload, the decreased cardiac output can also lead to severe bradycardia, exacerbating angina, heart failure, and pulmonary hypertension. Overdose can be treated by discontinuing the medication, chronotropic medications, and vasodilators [5].

There are no absolute contraindications for using this drug apart from hypersensitivity reactions such as anaphylaxis or less severe asthmatic episodes. Currently, no antidote is available to reverse this drug's effects. The treatment of hypertension and reflex bradycardia is discontinuing the administration of the drug [5].

2.1.1.1.2 Midodrine

Midodrine is a prodrug (medication that turns into active form once it enters the body). It is used to manage patients with orthostatic hypotension or hypotension secondary to other clinical conditions or drug therapies [7].

The chemical formula of this drug is C12H19ClN2O4. It is water-soluble and distributed as tablets for oral administration. Dosage forms are 2.5 mg, 5 mg, and 10 mg. Midodrine is almost completely absorbed after oral administration and undergoes enzymatic hydrolysis to form its pharmacologically active metabolite, de-glymidodrine. The drug should be stored in an airtight container [8].

The plasma levels of this prodrug peak at about half an hour and decline with a half-life of approximately 25 minutes. The peak concentration of the metabolites reaches about 1–2 hours, and their half-life is about 3–4 hours. The absolute bioavailability is 93% and is not affected by food. Midodrine deglycination to desglymidodrine appears in many tissues, and the liver metabolizes both compounds [9].

It does not act on cardiac β-adrenergic receptors and poorly diffuses across the blood-brain barrier. Increased embryo reabsorption is revealed in animal studies, as well as reduced fetal body weight and decreased fetal survival. There is no controlled data on human pregnancy. It is labeled US FDA Pregnancy category C, but the potential benefits may warrant the use in pregnant women, despite potential risks. No data is available for excretion in animal and human milk, but use should be avoided, and caution is recommended [8].

The contraindications to the drug are allergy to the drug, kidney disease, or, if one cannot urinate, pheochromocytoma (adrenal gland tumor), overactive thyroid, high blood pressure even while lying down, and liver disease. Taking this drug alongside other drugs that constrict the blood vessels can increase blood pressure. Common adverse effects (1–10%) of the drug are supine hypertension, paresthesia, headache, npiloerection, dysuria, nausea, dyspepsia, and vomiting [8].

The oral lethal dose (LD 50) is approximately 30–50 mg/kg in rats, 67.5 mg/kg in mice, and 125–160 mg/kg in dogs. Overdose symptoms could include hypertension, piloerection (goosebumps), a sensation of coldness, and urinary retention. The single doses associated with overdosage or potentially life-threatening symptoms in humans are unknown. Desglymidodrine is dialyzable [9].

2.1.1.2 Indirect-acting α-agonists

2.1.1.2.1 Ephedrine

Ephedrine is an α- and β-adrenergic agonist; however, it also causes the indirect release of norepinephrine from sympathetic neurons, inhibiting norepinephrine reuptake and displacing more norepinephrine from storage vesicles. Its use is indicated in treating hypotension under anesthesia, allergic conditions, bronchial asthma, and nasal congestion. Its chemical formula is $C_{10}H_{15}NO$ [10].

Ephedrine can be administered through oral, nasal, and intravenous routes (tablet/capsule:8–25 mg, Solution—0.5%, IV: 10–15 mg/1 mL). Ephedrine increases blood pressure by stimulating heart rate and cardiac output and variably increasing peripheral resistance. Activation of β-adrenergic receptors in the lungs causes bronchodilation. By stimulating α-adrenergic receptors in bladder smooth muscle cells, ephedrine also increases the resistance to the outflow of urine. Compared to when ephedrine is used to treat hypotension, using ephedrine for hypotension prophylaxis is associated with a higher risk of hypertension [10].

The bioavailability of ephedrine is 88%, and oral ephedrine reaches an average maximum concentration (C_{max}) of 79.5 ng/mL, with a time-to-peak concentration (T_{max}) of 1.81 hours [10].

Ephedrine is largely unmetabolized in the body and can be N-demethylated to norephedrine or demethylated and deaminized to benzoic acid conjugates and 1,2-hydroxypropyl benzene. The route of elimination is through the urine; about 60% is excreted as unmetabolized parent compound and 13% as benzoic acid conjugates and 1% as 1,2-dihydroxypropylbenzene. There is a large degree of inter-patient variability on the half-life of this drug, but orally, the plasma elimination half-life is approximately 6 hours [10].

Its use is contraindicated in people with cardiovascular disease, hypertension, hyperthyroidism, pheochromocytoma, and closed-angle glaucoma [11]. Large doses of ephedrine cause nervousness, insomnia, vertigo, headache, tachycardia, palpitation, and sweating. Some patients have nausea, vomiting, and anorexia. Painful urination may occur as a result of a vesical sphincter spasm. Urinary retention may develop in males with prostatism. Cardiac arrhythmias and precordial pain may occur following administration of ephedrine sulfate injection, USP [11].

The LD50 in mice after oral administration is 785 mg/kg, after intraperitoneal administration is 248 mg/kg, and after subcutaneous administration is 425 mg/kg. An overdose of ephedrine will present with rapidly increasing blood pressure. The overdose can be managed with blood pressure monitoring and possibly administering parenteral antihypertensives [10].

2.1.1.2.2 Methamphetamine

It is a sympathomimetic agent widely used to treat attention deficit hyperactivity disorder (ADHD) and exogenous obesity. Its chemical formula is $C_{10}H_{15}N$. Methamphetamine is a white solid odorless crystal. The recommended storage temperature is −20°C. This drug is a potent stimulant of the central nervous system, and it affects the neurochemical mechanisms responsible for regulating body temperature, heart rate, blood pressure, appetite, attention, mood, and responses associated with alertness or alarm conditions. The drug's acute effects closely resemble the psychological and physiological effects of an epinephrine-provoked flight-or-fight response; these responses include increased heart rate, vasoconstriction, increased blood pressure, hyperglycemia, and bronchodilation. It causes the elimination of fatigue, increased mental alertness, increased focus, and decreased appetite [12].

When methamphetamine enters the brain, it causes a cascade of norepinephrine, dopamine, and serotonin release. It acts as a dopaminergic and adrenergic reuptake inhibitor to a lesser extent, and in a higher concentration, it can act as a monoamine oxidase inhibitor [12].

Absorption of methamphetamine occurs in the gastrointestinal tract, with peak concentrations occurring at 3.13–6.3 hours after ingestion, and its effects may continue up to 24 hours in larger doses. When the drug is administered by inhalation, or intranasally, a high degree of absorption occurs. The drug has a high lipophilicity; it is distributed across the blood-brain barrier and crosses the placenta. This drug should be avoided in breastfeeding mothers as it is excreted through milk. The drug excretion occurs through the urine and increases with the acidic pH metabolization of methamphetamine occurs in the liver by aromatic hydroxylation, N-dealkylation, and deamination; at least seven metabolites have been identified in urine [12, 13].

The concurrent use of monoamine oxidase inhibitors with methamphetamine is contraindicated as a hypertensive crisis may occur. It is also contraindicated in patients with glaucoma, advanced arteriosclerosis, symptomatic cardiovascular disease, moderate to severe hypertension, hyperthyroidism, known hypersensitivity, or idiosyncrasy to sympathomimetic amines. It is a Pregnancy category C drug. It is shown to have teratogenic and embryocidal effects in mammals given multiple high human doses. There are no adequate and well-controlled studies in pregnant women, but it is recommended not to be used during pregnancy unless the potential benefit justifies the potential risk to the fetus [14].

Acute overdose of methamphetamine is manifested by restlessness, tremor, hyperreflexia, rapid respiration, confusion, assaultiveness, hallucinations, panic states, hyperpyrexia, and rhabdomyolysis. Fatigue and depression usually follow the central stimulation. Cardiovascular effects include arrhythmias, hypertension or hypotension, and circulatory collapse. Gastrointestinal symptoms include nausea, vomiting, diarrhea, and abdominal cramps. Fatal poisoning usually terminates in convulsions and coma [12].

Therapeutic methamphetamine blood concentration is 20–60 ug/dL; toxic methamphetamine blood concentration is 60–500 ug/dL, and lethal methamphetamine blood concentration is 1–4 mg/dL [12]. Benzodiazepines represent first-line treatment for methamphetamine toxicity but frequently require repeated and escalated dosing to achieve the effect [15].

2.1.2 α-2 receptor agonists

The α-2 receptors constitute a family of G-protein-coupled receptors (GPCRs) with three pharmacological subtypes, α-2A, α-2B, and α-2C [16]. Most α-2A and α-2C subtypes are located mainly in the presynaptic central nervous system. When stimulated, these receptor subtypes may be responsible for sedative, analgesic, and sympatholytic effects. On vascular smooth muscle, α-2B receptors are more prevalent and have been shown to mediate vasopressor effects. All three subtypes have been shown to inhibit adenylyl cyclase, resulting in decreased levels of cyclic adenosine monophosphate and hyperpolarization of noradrenergic neurons in the medial dorsal pons, specifically in the locus ceruleus [16]. As cyclic adenosine monophosphate is inhibited, potassium efflux via calcium-activated channels prevents calcium ions from entering the nerve terminal, inhibiting neural discharge. This process inhibits the release of norepinephrine and reduces the activity of ascending noradrenergic pathways, resulting in hypnosis and sedation. Activation of this negative feedback loop may also result in decreased heart rate and blood pressure, as well as a diminished sympathetic stress response. Stimulation of α-2 receptors in the spinal column's dorsal horn inhibits nociceptive neurons and reduces substance P release. Although there is evidence for supraspinal and peripheral sites of action, it is believed that the spinal mechanism accounts for the majority of the analgesic effects of α-2 agonist drugs [16].

Guanabenz, guanfacine, clonidine, tizanidine, medetomidine, and dexmedetomidine are all α-2 agonists with different potencies and affinities for different α-2 receptor subtypes. Clonidine, tizanidine, and dexmedetomidine have seen the most clinical use and will be discussed in greater depth.

2.1.2.1 Clonidine

Clonidine, an imidazole molecule, is a selective partial agonist for α-2 adrenoceptors with a 200:1 ratio (α_2–α_1).

Clonidine stimulates the brain stem's α-adrenoreceptors. This action diminishes central nervous system sympathetic outflow and decreases peripheral resistance, renal vascular resistance, heart rate, and blood pressure.

Clonidine can be administered *via* various routes: oral, intravenous, transdermal, rectal, and different neuraxial routes. It is rapidly and nearly completely absorbed following the oral route, reaching peak plasma levels in 60–90 minutes. A time-release transdermal patch can also administer clonidine; however, therapeutic levels require at least 2 days. It has an elimination half-life of 8–12 hours, with 50% of the drug metabolized in the liver to inactive metabolites and the rest being excreted unaltered in the kidney [17].

2.1.2.2 Clinical uses

Clonidine and guanfacine may be used to treat children and adolescents with attention deficit hyperactivity disorder. The reduced firing of presynaptic neurons releasing norepinephrine into the prefrontal cortex decreases the impulsive and hyperactive behavior seen in ADHD. Because of their additive effects on serotonin and γ-aminobutyric acid receptors, α-2 agonists are the most commonly utilized drugs to treat lack of sleep in children with ADHD. Clonidine also treats chronic pain disorders and withdrawal from opiates, benzodiazepines, alcohol, cocaine, food, and cigarette smoke [17].

Clonidine as an adjuvant has several advantages, including a reduction in the amount of opioids necessary for analgesia and hence a likely reduction in opioid-related side effects, titrated sedation and anxiolysis with no additive respiratory depression when combined with opioids and vasodilation and enhanced circulation of the cerebral, coronary, and visceral vascular beds [17].

Clonidine has lately been utilized as a premedication in individuals with considerable pretreatment anxiety. It has been found to improve mask application during anesthesia induction and to reduce anesthetic requirements by 40–60% in the pediatric population.

2.1.2.3 Dexmedetomidine

Dexmedetomidine, as clonidine, is a highly selective α-2 agonist with a higher affinity for the α-2 receptor (**Figure 1**). Clonidine has a specificity of 220: 1 (α-2: α-1), while dexmedetomidine has a specificity of 1620: 1. It is a full agonist of α-2 adrenergic receptors and the pharmacologically active d-isomer of medetomidine [18].

Dexmedetomidine induces a state of unconsciousness equivalent to normal sleep by activating central pre- and postsynaptic α-2 receptors in the locus ceruleus, with the added benefit of patients remaining easily stimulated and cooperative.

Dexmedetomidine generates a dose-dependent biphasic blood pressure response. Low-dose intravenous infusion lowers mean arterial pressure due to selectivity for central and peripheral α-2 receptors. The subsequent decreases in heart rate and systemic vascular resistance indirectly diminish cardiac output and systolic blood pressure. These actions help modulate the stress response, improve stability, and guard against drastic changes in cardiovascular parameters after surgery [18].

Dexmedetomidine can be given orally, intravenously, intramuscularly, buccally, and intranasally. It has a two-compartment distribution and elimination model. It has a ($T_{1/2}$ β) of 2 hours for elimination. However, it is a highly lipophilic medication that rapidly dispersed and redistributed, with a ($T_{1/2}$ α) of only 6 minutes for distribution. This has a

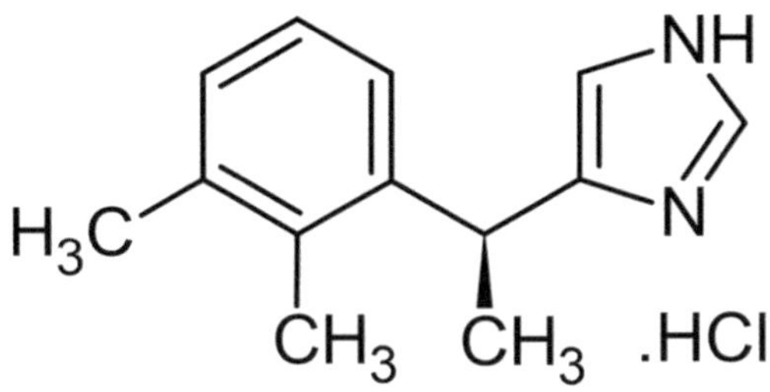

Figure 1.
Dexmedetomidine molecule.

rapid onset but only a short duration of clinical action. Because of its fast redistribution and removal, it is a suitable agent for infusion procedures. Dexmedetomidine is metabolized *via* direct glucuronidation and CYP2A6. Approximately 80–90% is eliminated in the urine, with the remaining 5–13% detected in the feces [18].

Pharmacokinetic interactions are unusual in most cases. However, dosage adjustments for concurrently administered sedatives may be required due to drug potentiation. Adding an α-2 agonist to a sedative regimen reduces the need for opioids by 50–75% and benzodiazepines by up to 80%. Dexmedetomidine's context-sensitive half-life ranges from 4 minutes after a 10-minute infusion to 250 minutes after an 8-hour infusion [18].

2.1.2.4 Clinical uses

Dexmedetomidine has three primary therapeutic applications: (a) in-hospital prolonged sedation, (b) procedure sedation and general anesthesia, and (c) obtunding emerging delirium. It is utilized as a sedative drug in critical care settings for critically ill patients who require prolonged sedation and mechanical ventilation. Dexmedetomidine possesses all of the features of an ideal critical care sedative. It does not cause respiratory depression, is analgesic and anxiolytic, has a fast onset, is titratable, and promotes drowsiness while maintaining hemodynamic stability. Finally, dexmedetomidine is exceptionally effective in treating the emerging delirium that can occur after general anesthesia, particularly in children. It has a significant relaxing effect without causing respiratory depression. This is a significant benefit over other medications typically used in such situations and requires additional study.

2.1.3 β-1 agonists

2.1.3.1 Dobutamine

Dobutamine is a synthetic sympathomimetic drug that selectively stimulates β-1 adrenergic receptors. It mimics the action of endogenous catecholamines like epinephrine but has a more specific effect on β-1 receptors. Dobutamine is typically available as a solution for intravenous infusion. The healthcare provider determines the concentration and dosage regimen based on the individual's specific needs.

Pharmacokinetically, dobutamine is administered intravenously due to its poor oral bioavailability. It has a rapid onset of action and a short duration of action. The drug is metabolized in the liver and excreted primarily in the urine.

Dobutamine acts primarily as a β-1 adrenergic receptor agonist. It increases the contractility of the heart muscle (positive inotropic effect) and enhances cardiac output. It also leads to mild vasodilation, primarily affecting the arterial system. In the clinical settings, dobutamine primarily treats acute heart failure or cardiogenic shock. It is used to improve cardiac contractility and increase cardiac output in these conditions.

2.1.3.2 Dopamine

Dopamine is a neurotransmitter and a sympathomimetic drug that acts on dopamine receptors in the central and peripheral nervous systems. It is crucial in physiological processes, including movement, motivation, reward, and blood pressure regulation. Dopamine is typically available as a solution for intravenous infusion. The healthcare provider determines the concentration and dosage regimen based on the individual's specific needs.

Pharmacokinetically, dopamine is administered intravenously due to its poor oral bioavailability. It has a rapid onset of action and a short duration of action. The drug is rapidly metabolized in the liver and excreted in the urine.

Dopamine acts on different receptors, including dopamine receptors, α-1 adrenergic receptors, and β-adrenergic receptors. Its effects vary depending on the dose administered. Dopamine primarily stimulates dopamine receptors at low doses, leading to renal and mesenteric vasodilation. It activates higher doses of α-1 and β-1 adrenergic receptors, increasing cardiac contractility and vasoconstriction.

Dopamine treats various conditions, including hypotension, shock, and low cardiac output states. It helps increase blood pressure and cardiac output by improving cardiac contractility and causing peripheral vasoconstriction.

2.1.3.3 Epinephrine (adrenaline)

Epinephrine, also known as adrenaline, is a naturally occurring catecholamine and a potent nonselective adrenergic agonist. It acts on α- and β-adrenergic receptors, producing various physiological effects. Epinephrine is available in different formulations, including solutions for intravenous injection, autoinjectors, and inhalers. The concentration and specific formulation may vary depending on the intended use.

Epinephrine can be administered *via* various routes, including intravenous, intramuscular, subcutaneous, and inhalation. It has a rapid onset of action and a short duration of action. The drug is metabolized in the liver and other tissues, and the metabolites are excreted in the urine. Epinephrine acts on both α- and β-adrenergic receptors. It produces various effects, including increased heart rate and contractility, bronchodilation, peripheral vasoconstriction, and increased blood pressure. These effects are beneficial in emergencies such as anaphylaxis, cardiac arrest, and severe asthma exacerbations.

Epinephrine is used in various emergencies, including anaphylaxis (severe allergic reactions), cardiac arrest, bronchospasm, and severe asthma exacerbations. It also restores blood pressure and maintains organ perfusion during resuscitation efforts.

2.1.4 β-2 agonists

β-adrenergic receptor agonists have long been used to treat both acute asthma symptoms and the prevention of exercise-induced asthma in adults and children and to treat COPD. They mimic the actions of catecholamines such as epinephrine, norepinephrine, and dopamine in triggering various autonomic responses within the body. β-2 agonists significantly affect the smooth muscle of the airway, uterus, gut, and systemic vasculature.

As part of our functional autonomic system, circulating catecholamines stimulate adrenergic receptors, resulting in parasympathetic and sympathetic physiological reactions. β-2 agonists operate as ligands to adrenergic receptors with higher selectivity for β-2 adrenergic receptors, mimicking catecholamines. When the β-2 adrenergic receptor is activated, a transmembrane signal cascade is initiated that includes the heterotrimeric G protein, Gs, and the effector, adenylyl cyclase. Adenylyl cyclase then raises intracellular cAMP through ATP hydrolysis. The increased cAMP concentration activates the cAMP-dependent protein kinase A (PKA). PKA can phosphorylate intracellular substrates, which modulate various actions within the cell. PKA, in particular, operates in airway smooth muscle to phosphorylate Gq-coupled receptors, resulting in a cascade of intracellular signals that have been postulated to diminish intracellular Ca2+ or decrease Ca2+ sensitivity [19].

The increase in Ca2+ inhibits myosin light chain phosphorylation, which prevents airway smooth muscle contraction. This is the underlying mechanism of β-2 agonists, which boost bronchodilatory effects and are used to treat a variety of common respiratory disorders. There have been suggestions that β-2 agonists have anti-inflammatory effects within the airway smooth muscle by decreasing intercellular adhesion molecule-1, decreasing granulocyte-macrophage colony-stimulating factor release, and stabilizing mast cell degranulation by inhibiting multiple inflammatory pathways.[T16] The duration and start of the action of β-2 agonists influence their classification. The three categories are short-acting, long-acting, and, most recently, ultra-long-acting β-agonists.

2.1.4.1 Short-acting β-2 agonists

They are first-line drugs for treating acute asthma symptoms and exacerbations. They are also often used in treating COPD in conjunction with long-acting, inhaled corticosteroids or long-acting muscarinic agonists. These drugs are often administered through inhalation, either metered dosage, dry powder inhalation, or nebulization. Compared to alternative oral delivery, inhalation has a higher therapeutic benefit and fewer systemic side effects. This family includes Salbutamol (albuterol), Terbutaline, Levalbuterol, and Pirbuterol.

2.1.4.2 Salbutamol

Salbutamol absorption is highly dependent on both formulation and dosage as well as the way of delivery. A thorough description of the effect of delivery systems on salbutamol pharmacokinetics may occupy many book chapters and is beyond the scope of this piece; however, we will consider some of the key points. Salbutamol is usually administered *via* a compressed metered-dose inhaler with an immense buffer. This very efficient delivery mechanism assures good distribution, especially to small- to moderate-sized airways. It can, however, be inhaled using a dry powder inhaler, or nebulizer, orally or intravenously.

Salbutamol, a partial agonist, has the greatest bronchodilating action at low dosages. It binds to β2-adrenoceptors located on airway smooth muscle (ASM) throughout the airways. This binding causes a postsynaptic action on adenyl cyclase, resulting in the formation of intracellular cyclic AMP (cAMP) from ATP, which in turn stimulates other effector molecules, including cAMP-dependent protein kinase A (PKA) and nucleotide exchange factor, which work together to cause intracellular Ca2+ sequestration, resulting in ASM relaxation. Despite that salbutamol's primary function is bronchodilation, it also suppresses mast cell mediator release and tumor necrosis factor α (TNF) release from monocytes. It also enhances mucus production and clearance of the mucociliary tract. It has extensive effects across numerous organ systems as a sympathomimetic, and it causes dose-dependent tachycardia, hyperglycemia, hypokalemia, and tremor. The systemic metabolic effects inducing glycogen breakdown and concomitant insulin release (possibly stimulated by pancreatic β2 cells) combine to cause high blood sugar levels and serum hypokalemia, with the former occurring as a consequence of cellular sodium excretion and potassium influx (Na-K-ATPase pump).

This side effect is beneficial in the emergency treatment of hyperkalemia, where ongoing salbutamol administration can decrease serum potassium between 1 and 1.5 mmol/L. However, it can also have complications such as dose-related tremors (the salbutamol shakes'), and when combined with cardiac receptor stimulation, stimulation can lead to tachyarrhythmias. Tachyphylaxis to β-2 agonists arises as soon as 1 week after starting regular medication and is more apparent with β 2-agonist monotherapy.

2.1.4.3 Terbutaline

It is also a selective β-2 adrenoceptor agonist used to prevent and reverse bronchoconstriction. Approximately, its volume of distribution is about 1.6 L/kg. After 72 hours, an oral dose of terbutaline gets eliminated in the urine by 40%. Terbutaline sulfate conjugated was the most predominant metabolite in the urine. Terbutaline parenteral levels are 90% removed in the urine, with roughly 2/3 as the unaltered primary substance. In the feces, less than 1% of a terbutaline dose gets eliminated.

2.1.4.4 Long-acting β-2 agonist

They are commonly used in managing asthma and COPD patients, often combined with inhaled corticosteroids. There is evidence that combination therapy is more effective than monotherapy. They have a longer onset time than short-acting medications, with salmeterol having an onset period of up to 15 minutes and lasting at least 12 hours. The suggested route of administration is inhalation, as with short acting. They are typically used as a second-line treatment in asthma patients who have failed to get clinical relief with short-acting medications. Salmeterol and formoterol are the commonest drugs in this group.

2.1.4.5 Salmeterol

In asthmatic patients, a 50 μg dose of inhaled salmeterol powder reaches a C_{max} of 47.897 pg./mL, with a T_{max} of 0.240 h and an AUC of 156.041 pg./mL/h. The distribution volume of the main compartment is 177 L, and the distribution volume of the peripheral one is 3160 L. Salmeterol is 96% protein linked to albumin and α-1-acid glycoprotein in plasma. It is primarily processed by CYP3A4 to α-hydroxysalmeterol1, with little contribution from an unknown process to an O-dealkylated metabolite.

Salmeterol is removed in the feces at 57.4% and the urine at 23%. Only around 5% of the dosage is excreted in the urine as unaltered salmeterol.

2.1.4.6 Formoterol

It has a fast onset of action (about 2–3 minutes) and an extended duration of action (up to 12 hours). In asthmatic patients, long-acting β-agonists such as formoterol without accompanying inhaled corticosteroids should be avoided, as long-acting monotherapy has been linked to an increased risk of asthma-related fatalities. Its pulmonary bioavailability is estimated to be around 43% of the delivered dose, whereas total bioavailability in the body is approximately 60% of the supplied dose (since systemic bioavailability comprises absorption in the stomach). Following inhalation, formoterol is rapidly absorbed into the plasma. Formoterol T_{max} in healthy adults ranged from 0.167 to 0.5 hours. C_{max} and AUC were 22 pmol/L and 81 pmol.h/L after a single dosage of 10 mcg, respectively. T_{max} in asthmatic adults ranged from 0.58 to 1.97 hours. C_{max} and AUC_{0-12h} after a single dose of 10mcg were 22 pmol/L and 125 pmol.h/L, respectively; after several doses of 10 mcg, C_{max} and AUC0-12 h were 41 pmol/L and 226 pmol.h/L, correspondingly. Across normal dosing ranges, absorption appears to be dose proportionate. It is 34–38% binding to plasma protein. It is predominantly processed by direct glucuronidation of the primary drug and O-demethylation of the primary drug, followed by glucuronidation. Minor mechanisms include primary drug sulfate conjugation and primary drug deformylation followed by sulfate conjugation, albeit these minor pathways have not been completely studied.

2.1.4.7 Ultra-long-acting β-2 agonist

Ultra-long-acting medications provide the longest duration of action, up to 24 hours, with the added benefit of being a once-a-day therapeutic dosage. The FDA has approved Indacaterol as a maintenance medication for COPD patients in combination with other bronchodilators. Indacaterol can be taken as a dry powder with a 5-minute onset of action. Many different ultra-LABAs are now being researched, with the potential to increase compliance and efficiency over current asthma and COPD therapy choices. Indacaterol, Vilanterol, and Oladaerol are the drugs in this group.

2.1.4.8 Administration

Metered-dose inhalers, nebulizers, dry powder inhalers, orally, subcutaneously, or intravenously are the most common delivery methods for β-2 agonists. Inhalation is the primary mode of delivery for β-2 agonists in treating asthma and COPD. Inhalation concentrates the therapeutic impact of the medicine on the airway's smooth muscles while minimizing the drug's diffusion to the systemic circulation. There is no association between the therapeutic impact of inhaled β-2 agonists and their peak plasma levels. Oral β-2 agonists, which have been demonstrated to exacerbate systemic side effects, are used less commonly. Terbutaline can also be administered intravenously, intramuscularly, or orally.

2.1.4.9 Adverse effects

The most prevalent side effect of β-2 agonists is desensitizing the β-2 adrenergic receptor to the β-2 agonist. Because adrenergic receptors have comparable features, β-2 agonists can have an "off-target" effect by stimulating α-1, α-2, or β-1 receptors. β-2 agonists' most prevalent adverse effects include the cardiovascular, metabolic, or musculoskeletal systems. Because of the vasodilatory impact on peripheral vasculature and a concomitant decrease in cardiac venous return, mechanisms of compensation show as tachycardia is relatively prevalent, particularly in the first few weeks of treatment. According to several publications ranging from single case reports to case-control studies, cardiac toxicity in the form of arrhythmias, cardiomyopathy, and ischemia has been more strongly associated with earlier-generation β-2 agonists. β-2 agonists have been demonstrated to lower serum potassium levels by causing an inward influx of potassium into cells *via* an action on the membrane-bound Na/K-ATPase, which can lead to hypokalemia. β-2 agonists also accelerate glycogenolysis, which might result in unintentional increases in serum glucose. Musculoskeletal tremors are another possible side effect, which is more familiar with using oral β-2 agonists. The severity of these side effects is often related to factors such as the affinity of each β-2 agonist to its specific receptor and medication dosages. Several studies additionally discovered hypoxemia and hypercapnia to be aggravating variables for β-2 agonist cardiotoxicity.[T25]

2.2 Sympatholytics

2.2.1 α-blockers

Sympatholytic drugs inhibit the effects of catecholamines by acting on postsynaptic adrenergic receptors present in target organs or by inhibiting the synthesis and storage of the catecholamines. These drugs can be divided into two subtypes, selective and nonselective α-receptor blockers.

Nonselective α-receptor antagonists block both the α-1 receptors as well as α-2 receptors. Blocking α-1 receptor causes vasodilation, while α-2 receptor blockade will reduce the force of vasodilation due to increased release of Norepinephrine. These medications, such as pheochromocytoma, are widely used in patients with increased sympathetic activity.

Selective α-1 receptor blockers act on the receptors and cause vasodilation; therefore, they are widely used in patients with hypertension and cause smooth muscle relaxation, so they help manage benign prostate hyperplasia [20].

The mechanism of action of *α-2 receptor blockers* is not known, although, in principle, they are known to inhibit negative feedback of norepinephrine release by stimulating the norepinephrine system, and they inhibit the effects of norepinephrine on postsynaptic α-2 adrenoceptors [20, 21].

2.2.1.1 Nonselective α-receptor blockers

2.2.1.1.1 Phentolamine

It is a nonselective α-receptor blocker used mainly to diagnose pheochromocytoma and to control or prevent paroxysmal hypertension immediately before or during pheochromocytoma ectomy. It is used to reverse soft tissue anesthesia, such as the tongue and the lips, and the associated functional deficits resulting from an intraoral submucosal injection of a local anesthetic containing a vasoconstrictor.

The drug is available in injection forms from 0.235 mg/1 mL to 10 mg/1/mL. The chemical formula of the drug is $C_{17}H_{19}N_3O$. α-receptors are present in blood vessels;

when they are activated by phentolamine, the blood vessels widen as the muscles relax and therefore decrease blood pressure. This drug maintains long-acting chemical sympathectomy. Phentolamine also stimulates β-adrenergic receptors and therefore causes a positive inotropic and chronotropic effect on the heart and increases cardiac output.

Phentolamine is only about 20% as active after oral administration as after parenteral administration. About 10–13% of the drug is eliminated unchanged in the urine, while the fate of the rest of the drug is unknown. The Tmax is 30–60 minutes. After intravenous administration of the drug, the elimination half-life is 19 minutes; after oral administration, it is 5–7 hours.

Some common adverse effects of the drug are weakness, dizziness, flushing, orthostatic hypotension, and nasal congestion, which have been reported in patients receiving phentolamine. Adverse GI effects are common and include abdominal pain, nausea, vomiting, diarrhea, and exacerbation of peptic ulcer. Adverse cardiovascular effects include prolonged hypotension, tachycardia, cardiac arrhythmias, and angina, especially after parenteral administration. Myocardial infarction and cerebrovascular spasm or occlusion, usually associated with marked hypotension and a shock-like state, have been reported occasionally following parenteral administration of phentolamine. Deaths have occurred after IV administration of phentolamine for the diagnosis of pheochromocytoma.

No specific antidote is available for phentolamine toxicity; however, in shock-like conditions such as a dangerous decrease in blood pressure or other evidence of shock, the person should be treated promptly with supportive care, and IV norepinephrine infusion can be administered if necessary. Epinephrine should not be used as it can cause a paradoxical decrease in blood pressure [22].

The oral LD50's (mg/kg) in mice is 1000, and in rats, it is 1250. No teratogenic or embryotoxic effects were observed in the rat, mouse, or rabbit studies, and no adequate and well-controlled studies in pregnant women are available. If the potential benefit of phentolamine justifies the potential risk to the fetus, the drug can be used. Whether or not the drug is excreted in human milk is unknown. As many drugs are excreted through human milk and since there is potential for adverse reactions in nursing infants, a decision should be made whether or not to continue the drug, considering the importance of the drug to the mother [22].

2.2.1.2 Selective α-receptor blockers

2.2.1.2.1 Prazosin

Prazosin is an α-1 receptor blocker used to treat hypertension, and recently, many studies have evaluated the drug's benefits in controlling post-traumatic stress disorder symptoms and associated nightmares. Other members of this drug class include Doxazosin, Terazosin, Tamsulosin, and Alfuzosin. This effect likely occurs through the inhibition of adrenergic stimulation found in states of hyperarousal. As this agent does not negatively impact lung function, it can manage hypertension in chronic obstructive lung diseases [23].

The chemical formula of the drug is $C_{19}H_{21}N_5O_4$. The usual adult for hypertension is 1 mg orally 2 or 3 times a day, initially, and the maintenance dose is 1–20 mg orally per day in divided doses [24]. It can be used alone or alongside other blood pressure-lowering agents, including diuretics and β-adrenergic blocking agents. The decrease in blood pressure may occur in both standing and supine positions [23].

After administering the oral dose, the peak plasma level of the drug is reached by approximately 3 hours, and the half-life is about 2–3 hours. Prazosin is metabolized in the liver by demethylation and conjugation and is excreted mainly in the bile and feces. The clearance of the drug is decreased in people with congestive heart failure.

As the drug lowers blood pressure, it can cause a clinically significant decrease in cardiac output, heart rate, blood flow to the kidney, and glomerular filtration rate. The decrease in blood pressure may occur in both standing and supine positions [23]. Shock caused by low blood pressure should first be treated with volume expanders, and vasopressors should be used if deemed necessary. Renal function should be monitored and supported as needed [25].

The LD50 in humans is 285 μg/kg orally. Severe drowsiness and decreased reflex occurred with ingesting at least 50 mg of Prazosin. There was no fall in blood pressure, and the child recovered without complications. The drug is classified as a Pregnancy category C drug. There are no adequate studies for determining the drug's safety during pregnancy. Specific cases of emergent use for blood pressure control during pregnancy showed no effects on the fetus or neonate. As the drug is excreted in small amounts in breast milk, it should be used cautiously in breastfeeding mothers.

Avoid alcohol and licorice with the use of this drug. Its absorption is not affected by food. Acute symptomatic liver injury due to prazosin is rare, and severe hepatotoxicity must be rare if it occurs at all [26].

2.2.1.2.2 Tamsulosin

It is an α-1A and α-1B adrenergic receptor antagonist used to treat benign prostatic hyperplasia, ureteral stones, female voiding problems, and prostatitis. The chemical formula is $C_{20}H_{28}N_2O_5S$(R38). It is available in the form of tablets, and the dose for treatment of adult benign prostate hyperplasia is 0.4 mg orally once a day; the dose may be increased to 0.8 mg orally once a day in patients who fail to respond to 0.4 mg once a day within 2–4 weeks [27].

The most significant effect of this drug is in the bladder and prostate, where the α-1A and α-1B adrenergic receptors are most common. The drug's action leads to the relaxation of prostate and bladder muscles, allowing for better urinary flow. Tamsulosin binds to α-1A receptors 3.9–38 times more selectively than α-1B and 3–20 times more selectively than α-1D. A significant effect on urinary flow with a reduced incidence of adverse reactions like orthostatic hypotension is allowed through this selectivity [28].

Tamsulosin is absorbed 90% in patients who are fasting. Taking the drug with food increases the time to maximum concentration from 4 to 5 hours to 6–7 hours but increases bioavailability by 30% and maximum plasma concentration by 40–70% [28].

The drug is metabolized by cytochrome P450 (CYP) 3A4 and 2D6 in the liver, with some metabolism by other CYPs. There is a low rise in liver transaminases by tamsulosin, but clinically, apparent liver injury is rare [28].

The oral LD50 in rats is 650 mg/kg. In an overdose, the patients might have hypotension that should be managed supportively by lying supine, administering fluids, or if further progression occurs, vasopressors might be needed, and renal function should be closely monitored. As tamsulosin is highly protein-bound, dialysis does not assist in treating overdose [28].

Animal studies have not shown any fetal harm caused by tamsulosin, but this drug is not indicated for use in women. Tamsulosin is excreted in the milk of rats, but no studies have been conducted about the effects of exposure to it. Male and female rats have been shown to have fertility affected by impairment of ejaculation and fertilization. In men, ejaculation problems have been recorded with the use of tamsulosin. At levels above the recommended dose, tamsulosin may be carcinogenic. There is a slight increase in mammary gland fibroadenomas and adenocarcinoma rates in female rats [28].

2.2.2 β-blockers

β-blockers block the physiological impacts of sympathetic nerve stimulation or circulating catecholamines on β-adrenoceptors, which exist across different organs in the body. Many organs have both β1 and β2 receptors coexisting (**Table 1**). For example, approximately 80% of the receptors are of the β-1 subtype in a typical individual heart. In heart failure, β1 receptors are downregulated, allowing a greater number of β2 receptors to be detected. The physiological and therapeutic effects of a β-blocker are determined by the actual quantity of β-1 or β-2 receptors in the various organs, the β-blocker's affinity, and the local drug concentration. When the bioavailability of β-blockers with a strong affinity for β-adrenoceptors is not too low, they can be helpful in small doses. Their effect persists even if they are washed out of the extracellular area. As a result, the plasma half-life of the β-phase of elimination cannot forecast their duration of activity. This is particularly true for many medicines that have a high affinity and a short plasma half-life (2–4 h for the β-phase).[T26]

Many β-blockers have additional features that may influence their value in individuals:

1. *Selectivity*: Considering β-blockers' desired effects are achieved by blocking β1-receptors, which dominate on the heart, "cardioselective" drugs with greater

Organ	Subtype	Function
Heart	β_1, $(\beta_2)^a$	Increase sinus rate Increase contractility Increase AV conduction
Gastrointestinal tract	β_1	Reduce muscular tone
Kidney	β_1, $(\beta_2)^a$	Increase Renin release
Fat cells	β_1, $(\beta_2)^a$	Increase lipolysis
Bronchi	β2	Reduce muscular tone
Blood vessels	β_2, $(\beta_1)^b$	Reduce muscular tone
Uterus	β_2	Reduce muscular tone
Pancreas (B-cell)	β_2, $(\beta_1)^a$	Increase insulin release
Thyroid gland	β_2, $(\beta_1)^a$	Increase T4 T3 conversion
Incretory glands	β_2 – $(\beta_1)^a$	Increase secretion of parathyroid hormone Reduce calcitonin & glucagon

[a]*receptor subtype coexistence.*
[b]*human cerebral blood vessels.*

Table 1.
Presence of β_1–β_2 receptors in various organ.

Selective β-1 receptor blockers	Non-selective β-blockers
• Metoprolol	• Carvedilol
• Atenolol	• Labetalol
• Nebivolol	• Propranolol
• Bisoprolol	• Satalol
• Acebutolol^{ISA}	• Timolol
• Betaxolol	
• Esmolol	

ISA: intrinsic sympathomimetic activity.

Table 2.
Selectivity of β-blockers.

sensitivity for this receptor are often recommended. However, "cardioselectivity" is not 100% and diminishes with increasing doses. Atenolol, bisoprolol, and metoprolol are examples of "cardioselective" β-blockers (**Table 2**).

2. *Partial agonist activity (intrinsic sympathomimetic activity)*: When baseline adrenergic firing is low (as during sleep), this manifests as a β-stimulant effect, but when adrenergic action is high (as during exercise), this manifests as β-blockade. Pindolol is a β-blocker with partial agonist action.

3. *Membrane-stabilizing activity*: Sotalol confers a local anesthetic and anti-arrhythmic effect.

4. *Additional characteristics*: Some β-blockers, such as carvedilol and labetalol, oppose effects conveyed at peripheral α-adrenoceptors, activate β2-adrenoceptors (e.g., celiprolol), or exhibit direct vasodilation effect (e.g., nebivolol).

The extent to which β-blockers are eliminated by the kidney or the liver varies, usually with considerable first-pass metabolism. Lipid-soluble β-blockers, such as labetalol, metoprolol, pindolol, and propranolol, are typically eliminated *via* the liver, whereas water-soluble β-blockers, such as atenolol, get eliminated by the kidney. The bioavailability of drugs removed by the liver varies significantly between populations. Most β-blockers have a short half-life; those removed through the kidney have a prolonged half-life.

2.2.2.1 Side effects

β-blockers have multiple unwarranted side effects secondary to their mechanism and site of actions, mainly:

1. Bronchoconstriction in susceptible individuals due to blockade of β-2 receptors, which mediate dilation in the bronchi. All β-blockers are contraindicated in the presence of asthma.

2. Bradycardia and cardiac contractility impairment.

3. Blockade of β-2 receptors, which serve vasodilation in blood arteries supplying skeletal muscle beds and cause peripheral vasoconstriction, resulting in cold hands and feet and possibly worsening Raynaud's phenomenon.

4. CNS symptoms related to diminished sympathetic discharge, such as malaise, intense dreams, nightmares, and, in rare cases, hallucinations, with highly lipid soluble β-blockers that have increased CNS penetration.

5. Restlessness and exhaustion are caused by β-2-receptor blockage in skeletal muscle, accompanied by increased muscular activity.

6. Hypoglycemia can be overlooked in insulin-dependent diabetes due to a reduction in sympathetic nerve stimulation.

3. Pharmacology of the parasympathetic nervous system (PANS)

The PANS, often called the "rest and digest" system, conserves energy and promotes homeostasis during periods of relaxation. Acetylcholine is the primary neurotransmitter in PANS signaling, acting on cholinergic receptors in various tissues. Cholinergic receptors are divided into two major types: nicotinic and muscarinic cholinergic receptors [1].

Nicotinic receptors are found at the neuromuscular junction and in the SANS and PANS ganglia. Activation of these receptors leads to subsequent muscle contraction or neurotransmitter release [1]. Nicotinic agonists, such as nicotine, are used primarily in smoking cessation therapies due to their stimulatory effects on the central nervous system [3]. In contrast, neuromuscular blocking agents, which act as nicotinic antagonists, are utilized in anesthesia to induce muscle relaxation during surgical procedures.

Muscarinic receptors: Muscarinic receptors are further classified into five subtypes, M1–M5. The PANS innervates these receptors in various target tissues, including the heart, smooth muscles, exocrine glands, and CNS structures. M1 receptors are predominantly located in the CNS, where their activation modulates cognitive function and memory. M2 receptors are primarily found in the heart, where activation slows heart rate and reduces contractility. M3 receptors are abundant in smooth muscles, glands, and endothelial cells. Stimulation of M3 receptors leads to bronchoconstriction, increased glandular secretions, and vasodilation.

The clinical utility of muscarinic agonists is limited compared to their antagonists. However, muscarinic antagonists, also known as anticholinergic drugs, play a crucial role in anesthesia. These agents, such as atropine and glycopyrrolate, counteract excessive PANS activity during anesthesia induction; prevent unwanted bradycardia, reduce salivary, and bronchial secretions; and facilitate intubation [1].

3.1 Parasympathomimetics

3.1.1 Muscarinic receptor agonist

3.1.1.1 Pilocarpine

Pilocarpine is a parasympathomimetic drug classified as a muscarinic receptor agonist. It is derived from the Pilocarpus plant and primarily acts on muscarinic receptors to produce pharmacological effects similar to acetylcholine. Pilocarpine is available in various formulations, including eye drops, tablets, and solutions. Eye drops are commonly used for ophthalmic purposes. Concentrations may vary depending on the specific indication. Storage conditions may involve protecting the drug from light and excessive heat.

Pilocarpine can be administered topically to the eye or orally. When applied topically to the eye, it has poor systemic absorption. Oral pilocarpine is well-absorbed from the gastrointestinal tract. The drug undergoes hepatic metabolism, primarily *via* hydrolysis, and is excreted mainly in the urine.

Pilocarpine selectively activates muscarinic receptors, predominantly the M3 subtype. It stimulates cholinergic receptors in various tissues, leading to miosis (pupillary constriction), increased salivation, sweating, bronchoconstriction, and gastrointestinal motility.

Clinically, pilocarpine eye drops are commonly used to treat glaucoma, where they reduce intraocular pressure by increasing the drainage of aqueous humor from the eye. Pilocarpine can also manage dry mouth (xerostomia) associated with Sjögren's syndrome or radiation therapy.

Pilocarpine is contraindicated in individuals with a known hypersensitivity to the drug, uncontrolled asthma, acute iritis, or narrow-angle glaucoma. It should be used cautiously in patients with cardiovascular diseases or gastrointestinal disorders. Other common side effects of pilocarpine may include localized ocular effects like temporary blurred vision, eye discomfort, or burning sensation when used as eye drops. Systemic effects can include increased sweating, increased salivation, gastrointestinal disturbances (such as nausea, vomiting, or diarrhea), and bronchoconstriction.

In cases of overdose or excessive use, pilocarpine can lead to excessive cholinergic stimulation. Symptoms may include profuse sweating, salivation, miosis, gastrointestinal distress, and potentially life-threatening cardiovascular effects. Treatment may involve discontinuing the drug, supportive measures, and administering atropine as a competitive antagonist to counteract the excessive muscarinic effects.

3.1.2 Acetyl-cholinesterase inhibitors

3.1.2.1 Neostigmine

Neostigmine is a reversible acetylcholinesterase inhibitor, classified as a parasympathomimetic drug. It increases the concentration of acetylcholine at cholinergic synapses by inhibiting the enzyme acetylcholinesterase, which breaks down acetylcholine. Neostigmine is available in various forms, including oral tablets and solutions for injection. The concentration and specific formulation may vary depending on the intended use.

Neostigmine can be administered orally, intramuscularly, or intravenously. It has poor oral bioavailability and is rapidly metabolized by esterases in the plasma and tissues. The elimination half-life is relatively short.

Neostigmine inhibits acetylcholinesterase, accumulating acetylcholine and exerting its effects at cholinergic synapses. It enhances neuromuscular transmission, leading to increased muscle strength and tone. It also affects cholinergic neurotransmission in other systems, such as the gastrointestinal tract.

Neostigmine is primarily used to manage myasthenia gravis, a neuromuscular disorder characterized by muscle weakness. It is also employed to reverse the effects of non-depolarizing neuromuscular blocking agents after surgery and to treat urinary retention.

Neostigmine is contraindicated in individuals with known hypersensitivity to the drug or those with mechanical gastrointestinal or urinary tract obstruction. It should be used cautiously in patients with asthma, epilepsy, or bradycardia. Other common side effects of neostigmine include gastrointestinal disturbances such as nausea, vomiting, diarrhea, and abdominal cramps. It may also cause increased salivation, sweating, bronchoconstriction, and bradycardia. These effects are related to its cholinergic activity.

In cases of overdose or excessive use of neostigmine, symptoms of cholinergic crisis may occur, including profuse salivation, sweating, bronchoconstriction, bradycardia, and potentially life-threatening respiratory depression. Treatment involves discontinuing the drug, administering atropine as a competitive antagonist, and supportive measures as necessary.

3.1.2.2 Physostigmine

Physostigmine is a reversible acetylcholinesterase inhibitor classified as a parasympathomimetic drug. It increases the concentration of acetylcholine at cholinergic synapses by inhibiting the enzyme acetylcholinesterase, which breaks down acetylcholine. Physostigmine is available in various forms, including oral tablets and solutions for injection. The concentration and specific formulation may vary depending on the intended use.

Physostigmine can be administered orally, intramuscularly, or intravenously. It is rapidly absorbed and metabolized by esterases in the plasma and tissues. The elimination half-life is relatively short. Physostigmine inhibits acetylcholinesterase, allowing acetylcholine to accumulate and exert its effects at cholinergic synapses. It enhances cholinergic neurotransmission in various systems, including the central nervous system and peripheral organs.

Physostigmine is primarily used to manage anticholinergic toxicity, including poisoning by anticholinergic drugs, such as certain medications, plants, or insecticides. It can reverse the effects of excessive anticholinergic activity, such as delirium, hallucinations, and peripheral manifestations. Physostigmine is contraindicated in individuals with known hypersensitivity to the drug or those with mechanical gastrointestinal or urinary tract obstruction. It should be used cautiously in patients with asthma, epilepsy, or bradycardia. Side effects of physostigmine include gastrointestinal disturbances such as nausea, vomiting, diarrhea, and abdominal cramps. It may also cause increased salivation, sweating, bronchoconstriction, and bradycardia. These effects are related to its cholinergic activity.

In cases of excessive use of physostigmine, symptoms of cholinergic crisis may occur, including profuse salivation, sweating, bronchoconstriction, bradycardia, and potentially life-threatening respiratory depression. Treatment involves discontinuing the drug, administering atropine as a competitive antagonist, and supportive measures as necessary.

3.1.2.3 Pyridostigmine

Pyridostigmine is a reversible acetylcholinesterase inhibitor classified as a parasympathomimetic drug. It increases the concentration of acetylcholine at cholinergic synapses by inhibiting the enzyme acetylcholinesterase, which breaks down acetylcholine. Pyridostigmine is available in various forms, including oral tablets and extended-release formulations. The concentration and specific formulation may vary depending on the intended use.

Pyridostigmine is primarily administered orally and is well-absorbed from the gastrointestinal tract. It has a more prolonged action duration than other acetylcholinesterase inhibitors, allowing for less frequent dosing. Pyridostigmine inhibits acetylcholinesterase, increasing acetylcholine concentration and enhancing cholinergic neurotransmission. It primarily acts on skeletal muscles, improving muscle strength and tone. It also affects cholinergic neurotransmission in other systems, such as the gastrointestinal tract.

Therapeutically, pyridostigmine is primarily used to manage myasthenia gravis, a neuromuscular disorder characterized by muscle weakness. It helps improve muscle strength and function in individuals with this condition.

Pyridostigmine is contraindicated in individuals with known hypersensitivity to the drug or those with mechanical gastrointestinal or urinary tract obstruction. It should be used cautiously in patients with asthma, epilepsy, or bradycardia. Side effects of pyridostigmine include gastrointestinal disturbances such as nausea, vomiting, diarrhea, and abdominal cramps. It may also cause increased salivation, sweating, bronchoconstriction, and bradycardia. These effects are related to its cholinergic activity.

Overuse of pyridostigmine can lead to symptoms of cholinergic crisis may occur, including profuse salivation, sweating, bronchoconstriction, bradycardia, and potentially life-threatening respiratory depression. Treatment involves discontinuing the drug, administering atropine as a competitive antagonist, and supportive measures as necessary.

3.1.2.4 Rivastigmine

Rivastigmine is a reversible acetylcholinesterase inhibitor classified as a parasympathomimetic drug. It increases the concentration of acetylcholine at cholinergic synapses by inhibiting the enzyme acetylcholinesterase, which breaks down acetylcholine. Rivastigmine is available in oral capsules, oral solutions, and transdermal patches. The capsules and oral solution come in various strengths, typically 1.5–6 mg, while the transdermal patches are available in different doses.

Rivastigmine can be administered orally or transdermally. When given orally, it is well-absorbed from the gastrointestinal tract. It undergoes extensive metabolism in the liver, and the elimination half-life varies depending on the individual's genetic makeup. Rivastigmine inhibits acetylcholinesterase, increasing acetylcholine concentration and enhancing cholinergic neurotransmission. It primarily acts in the central nervous system, specifically targeting acetylcholinesterase in the brain.

Rivastigmine is primarily used for the treatment of mild to moderate Alzheimer's disease and Parkinson's disease dementia. It helps improve cognitive function in individuals with these conditions, including memory, attention, and daily living activities.

Rivastigmine is contraindicated in individuals with a known hypersensitivity to the drug or those with a history of hypersensitivity to carbamate derivatives. It should be used cautiously in patients with gastrointestinal conditions such as peptic ulcer disease or those at risk of developing bradycardia. It shares similar side effects to its sister medications.

In cases of overdose or excessive use of rivastigmine, symptoms of cholinergic crisis may occur, including profuse salivation, sweating, bronchoconstriction, bradycardia, and potentially life-threatening respiratory depression. Treatment involves discontinuing the drug, administering atropine as a competitive antagonist, and supportive measures as necessary.

3.2 Parasympatholytics

Parasympatholytics are substances—or activities—that reduce the activity of the parasympathetic nervous system. They work by blocking the muscarinic receptors of the parasympathetic system. Most drugs with parasympatholytic properties are anticholinergics [29].

Parasympatholytic's pharmacodynamic effects include reduction of glandular secretion, dilatation of the pupil, paralysis of accommodation, increase of intraocular pressure, reduction of lacrimation, and more. These effects render parasympatholytics therapeutically valuable for treating slow heart rhythms, bronchioles constriction, and conditions such as benign prostatic hyperplasia, urinary retention, intestinal atony, and tachycardia [29]. It is worth mentioning, however, that parasympatholytics can interact with multiple drugs that can potentiate the antimuscarinic effect, such as antihistamines, neuroleptics, antidepressants, quinidine, or antiparkinson drugs [30].

Examples of parasympatholytics include atropine, methscopolamine bromide, flavoxate, orphenadrine, tiotropium, pinaverium, butylscopolamine, and anisodamine. However, atropine is the most used in the clinical setting.

3.2.1 Atropine

Atropine is classified as an anticholinergic or a parasympatholytic drug. Clinically, atropine is mainly indicated to treat bradyarrhythmias. Atropine also augments cardiac contractility by inhibiting cAMP-specific phosphodiesterase type 4, acting as a positive inotropic agent.

Atropine is a tropane alkaloid obtained from the deadly nightshade (*Atropa belladonna*) and other plants of the family Solanaceae. Its chemical formula is C17H23NO3, has a molecular weight of 289.4 g/mol (**Figure 2**), and is a racemic mixture of equimolar concentrations of (S)- and (R)-atropine. Atropine contains several functional groups, including an ester group, a hydroxyl group, and a tertiary amine group. The structure of atropine can be diagrammatically represented as benzene acetic acid, α-(hydroxymethyl)-8-methyl-azabicyclo {3.2. 1} oct-3-yl ester endo-(±). On hydrolysis, atropine gives (±)-tropic acid and tropine.

Atropine is an antimuscarinic agent that acts as a reversible, nonspecific antagonist of muscarinic receptors. It exerts its action by inhibiting the muscarinic actions of acetylcholine on structures innervated by postganglionic cholinergic nerves and smooth muscles, which respond to endogenous acetylcholine but are not so innervated. Atropine leads to both increased respiratory rate and depth, possibly due to the drug-induced inhibition of the vagus nerve. Generally, atropine counteracts the "rest and digest" activity of glands regulated by the parasympathetic nervous system.

Common medical uses of atropine include its role as an antisialagogue during surgery and anesthesia. It is also available in eye drops to treat uveitis and early amblyopia. Outside medicine, atropine is also used in the agricultural domain as a pesticide.

The pharmacological effects of atropine are due to binding to muscarinic acetylcholine receptors. Atropine is a competitive, reversible antagonist at muscarinic receptors, which blocks the effects of acetylcholine and other choline esters. Hence,

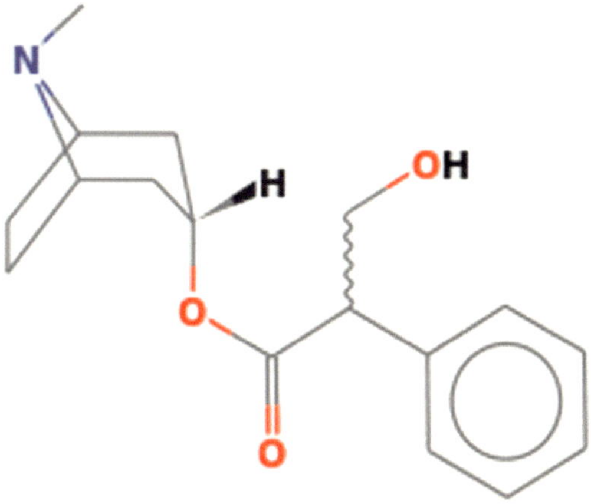

Figure 2.
Atropine molecule.

atropine is used as an antidote for poisoning by muscarinic agents, including organophosphates and other drugs.

Atropine can cause several side effects, mild or severe, depending on the dose and the individual's response to the drug. Some of the most common side effects of atropine include dry mouth, blurred vision, dry eyes, photophobia, confusion, headache, dizziness, fatigue, tachycardia, palpitations, flushing, urinary hesitance or retention, constipation, nausea, vomiting, and so on.

4. Conclusion

In conclusion, exploring autonomic nervous system (ANS) pharmacology presented in this chapter provides a comprehensive understanding of the intricate interplay between neurotransmitters, receptors, and drugs within the sympathetic and parasympathetic divisions. This chapter unveils the complexity of ANS modulation by dissecting the mechanisms of both agonists and antagonists and delving into direct and indirect drug actions.

The broad spectrum of examples discussed underscores the significance of ANS pharmacology across various medical disciplines. From managing hypotension and other medical problems, the clinical applications are far-reaching. The meticulous analysis of pharmacokinetics, metabolism, adverse effects, and contraindications empowers healthcare professionals to make informed decisions that optimize patient care.

For anesthesiologists, in particular, this knowledge is indispensable. The ability to finely tune autonomic responses during procedures can significantly impact patient outcomes and safety. A robust understanding of ANS pharmacology is a cornerstone of any physician toolkit, enabling them to navigate the intricate balance of autonomic control in the perioperative setting.

Conflict of interest

All authors declare no conflict of interest.

Author details

Redha Waseem[1], Mogahed Ismail Hassan Hussein[2], Tayseer Salih Mohamed Salih[2] and Sohel Mohamed Gamal Ahmed[2*]

1 Medical Education Department, Hamad Medical Corporation, Doha, Qatar

2 Department of Anaesthesiology, Intensive Care and Perioperative Medicine, Hamad Medical Corporation, Doha, Qatar

*Address all correspondence to: sohelm@yahoo.com

References

[1] McCorry LK. Physiology of the autonomic nervous system. American Journal of Pharmaceutical Education. 2007;**71**(4):78. DOI: 10.5688/aj710478

[2] Bankenahally R, Da M, Fcai F, Krovvidi H, Frca MM. Autonomic Nervous System: Anatomy, Physiology, and Relevance in Anaesthesia and Critical Care Medicine. Amsterdam, The Netherlands: BJA Education, Elsevier; 2016

[3] Clar DT, Sharma S. Autonomic pharmacology. In: StatPearls. Treasure Island (FL): StatPearls Publishing; 2023

[4] Bertram G. Katzung SBMAJT. Basic and Clinical Pharmacology. Vol. 12th Edition. USA: The McGraw-Hill Companies, Inc; 2012. ISBN: 978-0-07-176402-5

[5] Richards E, Lopez MJ, Maani CV. Phenylephrine. xPharm: The Comprehensive Pharmacology Reference [Internet]. 2022, 1-5. Available from: https://www.ncbi.nlm.nih.gov/books/NBK534801/ [Accessed: August 6, 2023]

[6] Phenylephrine: Uses, Interactions, Mechanism of Action | DrugBank Online [Internet]. Available from: https://go.drugbank.com/drugs/DB00388 [Accessed: August 6, 2023]

[7] Ali AA, Al-Ghobashy MA, Farid SF, Kassem MA. Development and validation of–MS/MS assay for the determination of the prodrug Midodrine and its active metabolite Desglymidodrine in plasma of ascitic patients: Application to individualized therapy and comparative pharmacokinetics. Journal of Chromatography B. 2015;**991**:34-40

[8] Midodrine Uses, Side Effects & Warnings [Internet]. Available from: https://www.drugs.com/mtm/midodrine.html [Accessed: August 6, 2023]

[9] Midodrine | C12H18N2O4 | CID 4195—PubChem [Internet]. Available from: https://pubchem.ncbi.nlm.nih.gov/compound/Midodrine#section=Absorption-Distribution-and-Excretion [Accessed: August 6, 2023]

[10] Ephedrine | C10H15ON | CID 9294—PubChem [Internet]. Available from: https://pubchem.ncbi.nlm.nih.gov/compound/9294 [Accessed: August 6, 2023]

[11] Ephedrine (Ephedrine): Uses, Dosage, Side Effects, Interactions, Warning [Internet]. Available from: https://www.rxlist.com/ephedrine-drug.htm [Accessed: August 6, 2023]

[12] Metamfetamine: Uses, Interactions, Mechanism of Action | DrugBank Online [Internet]. Available from: https://go.drugbank.com/drugs/DB01577 [Accessed: August 6, 2023]

[13] Methamphetamine | C10H15N | CID 10836—PubChem [Internet]. Available from: https://pubchem.ncbi.nlm.nih.gov/compound/10836 [Accessed: August 6, 2023]

[14] Desoxyn (Methamphetamine Hydrochloride): Uses, Dosage, Side Effects, Interactions, Warning [Internet]. Available from: https://www.rxlist.com/desoxyn-drug.htm [Accessed: August 6, 2023]

[15] Richards JR, Laurin EG. Methamphetamine toxicity. Pediatric Emergency Care. 2023;**15**(4):306. Available from: https://www.ncbi.nlm.nih.gov/books/NBK430895/ [Accessed: August 6, 2023]

[16] Giovannitti JA Jr, Thoms SM, Crawford JJ. α-2 adrenergic receptor agonists: A review of current clinical applications. Anesthesia Progress. 2015;**62**(1):31-39. DOI: 10.2344/0003-3006-62.1.31

[17] Jamadarkhana S, Gopal S. Clonidine in adults as a sedative agent in the intensive care unit. Journal of Anaesthesiology Clinical Pharmacology. 2010;**26**(4):439-445

[18] Lee S. Dexmedetomidine: Present and future directions. Korean Journal of Anesthesiology. 2019;**72**(4):323-330. DOI: 10.4097/kja.19259. Epub 2019 Jun 21

[19] Billington CK, Penn RB, Hall IP. β_2 Agonists. Handbook of Experimental Pharmacology. 2017;**237**:23-40. DOI: 10.1007/164_2016_64

[20] Miller SM, Cumpston KL. α Blockers. Encyclopedia of Toxicology: Third Edition [Internet]. 2020, 154-155. Available from: http://europepmc.org/books/NBK556066 [Accessed: August 6, 2023]

[21] Haller J, Makara GB, Pintér I, Gyertyán I, Egyed A. The mechanism of action of α 2 adrenoceptor blockers as revealed by effects on open field locomotion and escape reactions in the shuttle-box. Psychopharmacology. 1997;**134**(2):107-114. DOI: 10.1007/s002130050431

[22] Phentolamine Mesylate for injection, USP [Internet]. Available from: https://dailymed.nlm.nih.gov/dailymed/fda/fdaDrugXsl.cfm?setid=ed6c506c-5535-4b7c-ae1a-d63f09d796c9&type=display [Accessed: August 6, 2023]

[23] Prazosin: Uses, Interactions, Mechanism of Action | DrugBank Online [Internet]. Available from: https://go.drugbank.com/drugs/DB00457 [Accessed: August 6, 2023]

[24] Prazosin Uses, Side Effects & Warnings [Internet]. Available from: https://www.drugs.com/mtm/prazosin.html#dosage [Accessed: August 6, 2023]

[25] Minipress (prazosin) for High Blood Pressure (Hypertension): Uses, Dosage, Side Effects, Interactions, Warnings [Internet]. Available from: https://www.rxlist.com/minipress-drug.htm [Accessed: August 6, 2023]

[26] Prazosin | C19H21N5O4 | CID 4893—PubChem [Internet]. Available from: https://pubchem.ncbi.nlm.nih.gov/compound/4893#section=Structures [Accessed: August 6, 2023]

[27] Tamsulosin Uses, Dosage & Side Effects—Drugs.com [Internet]. Available from: https://www.drugs.com/tamsulosin.html [Accessed: August 6, 2023]

[28] Tamsulosin | C20H28N2O5S | CID 129211—PubChem [Internet]. Available from: https://pubchem.ncbi.nlm.nih.gov/compound/129211 [Accessed: August 6, 2023]

[29] Wilhelm H. Disorders of the pupil. Handbook of Clinical Neurology. 2011;**102**:427-466

[30] Berdel D, Berg AV. Use of parasympatholytics. Zhonghua Minguo Guo Xiao Er Ke Yi Xue Hui Za Zhi. 1989;**30**(6):378-388

Section 2

Autonomic Influences on Pain and Cardiovascular Responses

Chapter 3

The Role of Autonomic Nervous System in Pain Chronicity

Dmitry Kruglov and Dermot McGuckin

Abstract

The role of the autonomic nervous system (ANS) in chronic pain (CP) and in its chronicity is considered secondary and reactive to the nociceptive processes in the somatic nervous system (SomNS). However, research and clinical data strongly suggest the opposite. The ANS is an ancient, complex and ample part of the nervous system. It serves and controls visceral organs and somatic tissues. The ANS takes part in all aspects of all types of pain and influences its mechanisms at both peripheral and central levels. In this chapter we bring together the evidence from biomedical disciplines and clinical practice to support an alternative theory which contradicts the traditional views on the subject. We also raise questions which require further research to consolidate facts, advance our knowledge and improve treatment strategies for CP. The importance of this topic is difficult to overestimate because of the significant impact of CP on society and the lack of understanding, efficient therapy or cure.

Keywords: autonomic nervous system, autonomic, sympathetic, parasympathetic, chronicity, pain, chronic pain, visceral pain, somatic pain

1. Introduction

Chronic pain (CP) burdens a significant proportion of the population with pooled estimates for prevalence of 18–43% worldwide [1–3]. The vital role of the somatic nervous system (SomNS) in all types of pain is well recognised, but the importance of the autonomic nervous system (ANS) is mainly acknowledged in visceral or in 'sympathetically mediated' pain. The SomNS is perceived to be involved in all CP mechanisms and major dimensions of pain: physiological, sensory, affective, cognitive, behavioural, and sociocultural [4]. This is also true for the ANS, which is involved in major pain mechanisms and domains of all types of pain, not only visceral. A complete profile of ANS capability in CP formation has not been outlined, despite the accumulation of a sufficient body of evidence.

We planned this chapter as a brief conceptual narrative of a new notion of a comprehensive role of the ANS in CP. We will not didactically review the basic anatomy and physiology of the ANS, as we assume that anyone can find relevant information with its interpretation in current medical textbooks. Unfortunately, sometimes textbook authors present a simplified version of ANS structure and function. Without challenge, these deeply rooted views have been propagated from edition to edition or

Traditional views	Our suggestions
Primary role in development of musculoskeletal, neuropathic, and visceral chronic pain belongs to SomNS. Visceral CP is mainly sensed via somatic structures by the 'referred' mechanism. No CP develops without SomNS participation. The ANS plays only a secondary, reflective role in chronic pain.	The ANS plays a global role in CP, possibly more important than those of the SomNS. No CP develops without an essential ANS input. 'Referred pain' is only one of the mechanisms of visceral pain perception.
CP normally involves the sympathetic division of the ANS (e.g. CRPS, Fibromyalgia). The parasympathetic division of the ANS produces mostly an anti-inflammatory, anti-stress and, in general, a positive effect in CP conditions.	Sympathetic and parasympathetic divisions of the ANS have complex anatomical and physiological relationships, and both participate in CP development.

Table 1.
A brief summary of authors suggestions versus traditional views on the role of ANS in chronic pain development.

referenced in other publications. For example, it is widely considered that a leading role in CP development belongs to the SomNS; the ANS only responds to acute or already established CP. This conclusion frequently follows the outcomes of experimental studies [5]. We propose the opposite: the ANS plays a primary role in any type of CP and in pain chronicity. **Table 1** summarises and compares our suggestions with traditional beliefs on the subject:

Our view challenges current understanding of ANS involvement in pain chronicity and opens new avenues for diagnosis, treatments, and outcome monitoring. Our opinion draws on basic facts and advanced knowledge of different fields including, but not limited to, evolutionary biology, anatomy, epidemiology, pathophysiology, diagnostics and western and traditional medicine (i.e., effects of treatment). Therefore, the structure of this chapter follows the above list of biomedical disciplines and encompasses illustrative examples of medical treatment, interventions, and investigations. Surprisingly, there are still many gaps in CP theory. By highlighting them and asking appropriate questions we hope to encourage independent thinking and to stimulate future research. This is aimed to improve an evidenced based approach to refractory CP conditions which burden our society.

2. Evolutionary biology

Acute pain is one of the essential phenomena in biology because it helps organisms to survive. Therefore, it must have emerged early in phylogenesis, and since then it has evolved along with the growing complexity of the nervous system. Our knowledge about nervous system evolution lacks satisfying clarity [6]. For example, it is not clear when the central nervous system (CNS) appeared or whether it debuted independently more than once in the history of animal life on Earth [7]. Also, we do not know for certain if the SomNS arrived before the ANS, or whether the sympathetic division of the ANS developed earlier than the parasympathetic division. Some embryological studies report that the oldest autonomic structures were unmyelinated vagal fibres from the dorsal motor nucleus of the vagus [8].

Scientists face significant difficulties answering the above questions as nervous tissue does not preserve well in fossils. To support any of these conflicting points, authors sometimes use a 'common sense' approach by asking what is more important, ability to 'fight or flight', or control of the internal milieu in a precise way. One theory proposes that the "evolutionary origin of brainstem parasympathetic motor neurons out of branchial motor neurons, and spinal sympathetic motor neurons out of spinal motor neurons" [9].

However, comparison of ancient (but still living) species with modern organisms gives us essential facts for better understanding of evolutionary puzzles. For instance, sympathetic systems early in history employed acetylcholine (ACh) in postganglionic efferent neurons, and only later the majority of these fibres switched to noradrenaline (NA), except for sudomotor fibres. Certainly, the neuromediator change was reflected in sympathetic influence on some target organs with dual (sympathetic and parasympathetic) supply, when the stimulating effect mediated by ACh was passed over to the parasympathetic system.

ANS centres are located in phylogenetically ancient areas of the brain (hindbrain and midbrain) where autonomic, as well as the old somatic pain pathways (i.e., paleospinothalamic and archispinothalamic), terminate or make connections to; while the new somatosensory pain pathway (neospinothalamic) travels to the neocortex (forebrain).

Conclusion: SomNS and ANS have similar peripheral nociceptors, use the same neurotransmitters, and often share anatomical pathways and central connections. Both systems are of similar phylogenetic age [10, 11] and both should have been equally involved in pain processing, analysis, and responses.

3. Anatomy

Vast anatomical data help us to appreciate the fundamental role of the ANS in any type of CP. One can observe this in the complexity of the peripheral ANS: intricate structure and autonomy of local reflexes; abundancy (present in somatic and visceral peripheral nerves); diversity (variety of neuron types with different functions) [12]; and phylogenetic age of its spinal cord tracts and brain centres. Specificity in anatomical organisation of the ANS is the reason for precise homeostatic control: the efferent ANS (by its diversity of function and size) significantly "outweighs the somatic efferent pathways" [13]. Throughout this chapter we will continue comparing the ANS with the SomNS and draw your attention to their close interactions and inseparable activity. In this section we discuss the afferent, central, and efferent parts of the ANS and their significance for CP, but we will not cover the enteric part of the ANS.

3.1 Afferent ANS

Practically all somatic nerves contain autonomic fibres, and all of them are considered to be of efferent type (sudomotor, pilomotor and vasomotor). However, not all nerves contain somatic fibres; visceral nerves consist only of autonomic fibres. Therefore, any damage to peripheral nerve(s) will affect the performance of the peripheral ANS. This is also true for any type of damage at the level of the spinal cord as autonomic pain (sensory) tracts are in close proximity to somatic pathways. In consequence, when interventions treat peripheral nerves, nerve roots or epidural space, or spinal cord targets, the therapeutic effect on somatic and visceral/autonomic structures often cannot be differentiated.

Some publications advocate that primary visceral (sensory) neurons do not belong to the ANS, or if they do, they are not divided into sympathetic or parasympathetic fibres (despite visceral afferents travelling along sympathetic or parasympathetic nerves). This concept is often oversimplified in the literature. In order to understand the matter, one has to explore: the definition of the ANS; types of fibres in the peripheral nervous system (somatic and autonomic) and connections their primary sensory neurons make in the spinal cord; ascending tracts (with their targets, number of neurons involved and their functions).

Definitions of the ANS which are currently used by various dictionaries, institutions, publications, and other sources of information, typically declare that it controls involuntary functions of the body (internal organs and glands). Some interpretations might add a conflicting statement, for example, characterising the ANS as a part of the peripheral nervous system only (a network of peripheral nerves and ganglions), or suggest that the ANS has only motor or efferent fibres. This ambiguity undermines the functional intricacy and capacity of the ANS, which consists of various afferents, spinal cord and brain tracts and a network of analysing and executing centres.

Afferent innervation for internal organs comes from vagal (85% of vagal fibres) and spinal (50% of splanchnic nerves fibres) visceral afferent neurons [14]. The neuron cell bodies of these afferents are located in dorsal root ganglions (DRG) or in cranial nerve (IX and X) ganglions. Vagal afferent neuron projections are organised viscerotopically within the solitary tract in the medulla and spinal visceral afferent neurons connect to Rexed laminae I and V and deeper layers of the spinal cord in a segmental order [14].

Different types of neurons outside and inside CNS carry molecule and transcription factor signatures. Vagal afferent and efferent neurons, sympathetic post-ganglionic and autonomic neurons in the CNS are defined by homeodomain transcription factor *Phox2b* [15, 16], but visceral spinal afferents are not. The latter, in contrast to somatic sensory neurons, do not typically target Rexed lamina II but give rise to different pain pathways within the spinal cord.

General visceral afferents which travel with sympathetic peripheral nerves have their cell bodies in DRGs and their axons synapse with the second-order sensory neurons predominantly in laminae I and V as well as deeper laminae (VII, VIII and X) of the spinal grey matter. Spinal afferents which project to viscera comprise only a few percent of all sensory neurons in DRGs, the vast majority of neurons there are of somatic nature. Visceral afferents which ascend along vagal nerves have their cell bodies in inferior (nodosum) ganglions and some in superior (jugular) ganglions. They project to the nucleus tractus solitarius (NTS) of the brainstem. We are not going to discuss autonomic afferents of VII (facial) and IX (glossopharyngeal) cranial nerves here.

Spinal visceral afferents are not morphologically different from somatic afferent neurons with cell bodies in DRGs. However, they might differ by their spinal cord pathways and a number of additional functions. These are local efferent and trophic roles, both related to antidromic transport and release of chemicals and mediators via the afferent terminals to the innervated cells to influence visceral activity. Vagal visceral afferents show a great deal of diversity and coding strategies in respect to the organs these neurons serve [17–20]. The conventional view is that nociceptive information does not get transferred via fibres within the vagus nerve. However, there are data supporting participation of vagal afferents in pain directly and indirectly. The latter might include interaction with sympathetic afferents at the cervical level [21], by inhibition of nociceptive dorsal horn activity, or by mediation of unpleasant symptoms (like nausea and bloating) which can exacerbate the pain experience. It is important to note that pelvic organs receive afferent innervation from two sources, both lumbar and sacral outputs. This list of possible mechanisms of visceral afferents in pain transmission and modulation is not exhaustive. The majority of evidence is based on animal studies, but due to high level of phylogenetic conservation, the majority of anatomical and physiological data could be applied to humans.

3.2 Ascending spinal cord tracts and vagal projections

The complexity of the afferent part of ANS is not fully discovered. New research emerges every year clarifying some and giving start to new questions. However, the situation with ascending autonomic spinal cord pathways is even more perplexing. The confusion comes from traditional descriptions of the ascending sensory pathways, including:

- name and phylogenetic age of a particular tract and its position within spinal cord;
- number of neurons and synaptic connections involved;
- laterality (ipsilateral, contralateral, or bilateral);
- destination(s) and branches to other brain centres;
- communication to somatic sensory pathways;
- connection(s) to motor tracts (and reflex activity);
- descending modulating and inhibitory effects exerted by neurons of interest.

For the purpose of this chapter, we allocate the highest significance to destination of the tracts and interconnections to other pathways. The latter feature enhances sensory experience and responses, including neuromodulating functions. It would be also useful to pay attention to evolutional order of appearance. Here we are not going to talk in detail about anatomical position and laterality. We will briefly mention this information only for selected pathways, as it is important in relation to accessibility by pain relieving interventions (their successes or failures).

Studies of nociceptive ascending spinal cord pathways confirmed existence of a large group of tracts. Not all of them end up in the brain cortex, many relay information to various areas of phylogenetically older parts of brain. Activation of these areas together with cortex centres contributes to multidimensional pain experience and its chronicity.

Fibres making the shortest (oligosynaptic) way to the cerebral cortex belong to relatively young structures (found in higher mammals), hence, forming the lateral spinothalamic (neospinothalamic) tract. It is monosynaptic on the segment to thalamus and, therefore, the fastest one. It brings sharp and well-localised sensation (small receptive fields) with a definitive quality (burning, stinging etc) of various intensity. These signals reach out to the somatosensory cortex; therefore, alert and warn consciousness. Neospinothalamic fibres are somatotopically organised (in all connections and at all levels), crossing to the opposite side in the spinal cord, and carry only somatosensory (not visceral) nociceptive information.

The older parallel tracts, named paleospinothalamic and archispinothalamic, include one or more synapses before thalamus. They make extensive connections to brainstem and other brain structures, lack somatotopic arrangements, target internuclear thalamus nuclei, and start subconscious autonomic and descending neuromodulating reflexes. These pathways tap into the affective dimension of pain experience.

Autonomic nociceptive signals travel via older tracts. Primary autonomic sensory neurons converge their input on the next neuron together with somatosensory afferents. This happens in the grey matter of the spinal cord. Viscero-somatic wide dynamic range neurons take input from large diameter (myelinated skin afferents) and smaller (myelinated and non-myelinated skin and deep tissue afferents) and primary visceral afferents. Visceral, as well as somatic, nociceptive information (via convergent neurons) could be transmitted via multiple pathways:

- Spinobulbar—targeting Ventrolateral Reticular Formation (VRF), Dorsal Reticular Nucleus (DRt), Nucleus Tractus Solitarii (NTS), Rostral Ventromedial Medulla (RVM);

- Spinopontine—targeting most studied Parabrachial Nucleus (PBN);

- Spinomesencephalic—targeting Periaqueductal Grey (PAG);

- Spinodiencephalic—targeting nuclei of Lateral and Medial Thalamus, Hypothalamus.

Majority of the above anatomical discoveries were done with fine antero–/retrograde tracing techniques on animals (rats, cats, and monkeys) but often the results are transferrable to humans because of high level of phylogenetical conservation. The more recent studies revealed the presence of direct tracts connecting spinal cord with cortex and subcortical telencephalon bypassing thalamus.

The significance of simultaneous activation of parallel oligo- and polysynaptic ascending nociceptive pathways is not fully researched; however, we can appreciate its contribution to vivid reality of pain or pain relief in everyday life. This is also important prediction of outcome (and duration) of pain-relieving ablative procedures [22].

The above-mentioned ascending (relaying visceral nociceptive information) tracts project signals further by connecting to RVM, DRt, pontine noradrenergic groups, the hypothalamus, amygdala, the ventrolateral medulla VLM, the NTS, the rostral ventromedial medulla, PBN, the PAG, the thalamus and cortex (parietal somatosensory, prefrontal, frontal motor, orbital and cingulated). Majority of these destinations are parts of ANS. Some of the descending circuits which originate from PAG, DRt and some other centres exhibits suppression and facilitation at spinal cord synaptic sites of ascending tracts.

Visceral pain is also transferred by midline postsynaptic dorsal column (PSDC) pathway. The axons of PSDC neurons transmit pelvic visceral nociception, they travel uncrossed in the dorsal column. The primary termination of the visceral input of the PSDC cells is the dorsal column nucleus. Pelvic visceral cancer pain responds to limited or punctate surgical midline myelotomy, thoracic visceral pain—to a lesion at the lateral edge of the gracile fasciculus, and experimental pancreatic pain to complete bilateral lesion of the gracile fasciculus [23, 24].

3.3 Central ANS

As discussed earlier, ascending ANS spinal cord tracts project to many brain locations and, therefore, are capable of production of multiple effects including descending pain control. Although, autonomic nociceptive pathways do not directly influence somatic pain, both systems meet at the spinal cord level (when primary afferents converge). Non-discriminative somatosensory nociception shares (phylogenetically old) pathways with visceral afferents and, therefore, highlight the same areas of the brain. Central and, therefore, efferent parts of ANS might be involved in visceral and at the same time in somatic pain due to such overlap. This is also reflected in the fact that "Pain Matrix", network of brain centres responsible for pain processing shares key areas with the ANS.

The traditional take on these relationships is that the ANS passively responds to acute or chronic pain. Observers measure a shift of autonomic balance between sympathetic or parasympathetic tone using heart rate variability (HRV) or other tests (sudomotor activity, muscle sympathetic activity etc) or simply vital signs (heart rate, rate of breathing). We consider this topic is largely uncovered and deserves more attention.

The ANS governs body functions and influences mental state, emotions (conscious and subconscious phenomenon) and feelings (conscious phenomenon). Its centres include those of forebrain (insular and anterior cingulate cortex, amygdala, hypothalamus) and brainstem (PAG, PBN, NTS, VLM and some other parts of medulla). There is a fast-growing body of publications showing the complex relationship between pain conditions and activity of ANS centres [25]. There are many examples of this co-existence in people with diseases of various systems: cardiovascular [26]; respiratory [27, 28]; digestive [29, 30]; genitourinary [31, 32]; immune [33]; thermoregulation [34]; cerebral circulation and headaches [35–37]; sleep and circadian rhythms [38, 39]. Many of the above publications and similar are observational studies (and rarely prospective) or reviews. It is difficult to say if chronic pain was the reason for recorded changes or if pre-existing disturbances in ANS functions created vulnerability to chronic pain.

We would argue that in real life, disturbing and disabling chronic pain cannot develop without disorder of ANS control. Disturbed autonomic functions and chronic pain are in reciprocal relationship often with positive feedback: chronic pain might be a reason for autonomic symptoms and developed symptoms might reinforce and facilitate duration of pain, and they usually trigger and exacerbate each other. Therefore, frequently pre-existing autonomic derangement (even mild) due to lifestyle or any other reasons makes people more vulnerable to development of chronic pain.

3.4 Efferent ANS

In many sources, the ANS is considered a binary structure with sympathetic and parasympathetic divisions. For simplicity of teaching, the efferent output of the ANS is divided into craniosacral (parasympathetic) and thoracolumbar (sympathetic), with opposite effects on target organs. However, this is true only for a few targets; many internal organs, glands, skin structures and blood vessels are innervated only by one division. If both divisions are involved, they do not produce opposite responses (stimulation vs. suppression) or each branch functions under different conditions. So, the correct view is that ANS divisions work synergistically to provide stability of the internal environment and provide with the adaptive responses for internal organs and somatic structures.

The anatomy of the efferent ANS is more complex than that of the somatic motor system. It has preganglionic segments, ganglions, and postganglionic motor neurons. Detailed structure of the efferent ANS is well described in the literature. We will touch only upon its relevance for CP development and perpetuation.

A vital point supporting our view is based on the involuntary reflexes which are delivered by autonomic efferent fibres to visceral and somatic targets. The afferent information for this activity comes from visceral or somatic sources, but the central nuclei belong to the ANS. Physiological reflexes might change under pathological conditions as well as the homeostatic control of internal organ functions. This could lead to a variety of symptoms and painful conditions. These changes might replace the original programs and become chronic through learning and neuroplasticity mechanisms. Altered function (i.e., bowel contraction, acid production or abnormal blood supply) might cause more pain and unpleasant sensations perceived through the ANS.

Pain inherently boosts pathological neuroplasticity through reinforcement learning where the insula plays an important role [40, 41]. The underlying mechanisms could negatively affect physiological training-induced neuroplasticity in physical tasks [42]. However, the precise effect of pain (acute, experimental, or chronic) on motor skill learning is the subject of debate as no strong evidence has been provided by research [43]. The longer the duration of reflex changes, and of pain, the more complex the situation becomes and the more challenging and less successful treatment is. At some point, the pain condition reaches an irreversible phase [44], when treatment pursues palliative outcomes.

Examples of altered reflexes affecting visceral organs could be Irritable Bowel Disease (IBS), where pain is linked to abnormal gut motility, or urinary bladder conditions, when disturbing symptoms of urgency, incontinence and spasms convolute with pain. As for somatic organs affected by pathologically-changed autonomic reflexes, (e.g., skin thermoregulation) the afferent part is mediated by somatosensory afferents, but the central control (hypothalamus) and efferent output (sudomotor, vasomotor and pilomotor nerve fibres) is provided by the ANS [45].

Recent research has revealed an interesting relationship between autonomic and somatic neurons, which might explain muscle weakness in certain painful conditions with altered sympathetic outflow to muscles, for example CRPS. Sympathetic efferent fibres innervate neuromuscular junctions and are vital for maintenance and function of synapses between somatic motor nerves and muscles [46, 47].

In *The Senses: A Comprehensive Reference, Second Edition (2020)*, chapter 5–21 [48] possible means by which the sympathetic nervous system could influence CP in somatic tissues are summarised. These are as follows: sympathetic-somatic afferent coupling; sensitisation of somatic nociceptors; neurogenic inflammation; and central changes in sympathetic pathways with the release of a variety of neuroactive substances and participation of neuroendocrine system. Earlie, Prof Jänig [49] discussed sympathetic (other than sudo–/pilomotor or vasoconstrictor) innervation of skin and deep somatic tissues, including muscles (vasodilators) and bones (peptidergic neurons, probably affecting mineralisation).

ANS involvement in musculoskeletal pain was investigated on the model of delayed onset muscle soreness (DOMS). Fleckenstein *et al* set out to discover to what extent sympathetically mediated pain (SMP) is responsible for exercise-induced acute muscle pain or damage in the upper limb [50]. They found that sympathetic regional (stellate ganglion) blockade causes pain relieving and anti-inflammatory effects. They suggested mechanisms for these effects which outlasted local anaesthetic block. These could be due to interruption of the vicious cycle of pain and local reflexes [51], allowing a reboot or a change in cytokine profile from pro- to anti-inflammatory [52]. There is also a possibility of changing of sympathetic and parasympathetic balance secondary to regional sympathetic outflow interruption. In fact, these effects might follow any peripheral neural injection as autonomic fibres will always be affected by local anaesthetic due to the abundance of peripheral ANS fibres, as it was mentioned above.

Conclusion: Complexity (of all parts) and ample presence of the ANS; interconnections within and with the SomNS; active control of body functions (involved in pain mechanisms), emotions and behaviour; neuroplasticity of reflexes support evidence of global and fundamental role of the ANS in pain development and chronicity.

4. Epidemiology

Epidemiological studies have highlighted the prevalence of different pain conditions and their risk factors. Despite ongoing research, some of these facts (e.g., uneven gender distribution, drug sensitivity with therapeutic response, associations with other diseases) are difficult to explain. ANS imbalance might be one of the reasons we overlook.

Gender difference in pain prevalence, sensitivity and analgesic response has been reported in the literature [3, 53]. Women experience more severe pain and report it more frequently. They develop pain in more anatomical sites and for a longer time, with a higher prevalence across the majority of pain conditions. But some painful disorders are strikingly more frequently seen in women: fibromyalgia; pelvic and musculoskeletal pain; and temporo-mandibular joint pain amongst others. This is routinely attributed to genetic factors, sex hormone profile and cyclical changes in serum concentrations, tissue nerve density, and psychological factors, but rarely to ANS input.

Previous research has shown that women have a prevailing parasympathetic tone whereas men have a prevailing sympathetic tone. This difference disappears after the age of 55 years [54]. Sympathetic system activation has been reported in CP states. We do not know how the parasympathetic division contributes, but this certainly involves complex, multilevel and non-linear interrelationships between the sympathetic division and other determinants.

One of the conditions which is three to four times more common in females is complex regional pain syndrome (CRPS). Reported risk factors for CRPS include: history of migraine; osteoporosis; asthma; and angiotensin converting enzyme (ACE) inhibitor therapy. The latter two are associated with parasympathetic predominance, but osteoporosis is considered more related to sympathetic activation [55].

Neurotransmitters are used in experimental research to obtain strong evidence on the mode of activity of ANS structures in question. Drugs which we prescribe for treatment of any illness might intentionally (indication) or unintentionally (side-effect) shift the balance between autonomic divisions. This is mediated through direct or indirect effects on adrenal and acetylcholine receptors in peripheral or central ANS. For instance, many antihypertensives suppress sympathetic outflow, whilst some antidepressants block and some opioids stimulate cholinergic pathways. Thus, treatment of comorbidities might affect pain conditions and vice versa. This is also important when pain-relieving drugs are chosen for a particular individual.

Time course of chronic diseases corresponds to constant changes which the ANS undergoes due to ageing, adjustment to climate, food habits, physical activity and many other factors. For example, asthma is more prevalent in boys, but later in life becomes more prevalent in women. This probably is due to a shift in autonomic balance which affects ANS airway control.

Conclusion: When we assess a case of CP it is essential to understand how it is related to excess or insufficient activity in each ANS division. This might influence our choice of drugs, interventions and other treatment methods, as well as help prognosticate. Coexisting medical conditions might give us a clue about autonomic balance and its dynamics, however, future research with appropriate questions and a fresh view on the problem might shed more light on the matter. The role of the parasympathetic system in developing CP has not been fully elucidated.

5. Pathophysiology

Practice makes perfect. This maxim is fully applicable to ANS design and functioning, but it requires a few clarifying comments. The ANS functions according to inherited programs (reflexes), and by learned behaviour patterns, which are established and maintained since birth and childhood [56]. This is achieved through learning by continuous feedback from internal and external environments, and neuroplastic changes which strengthen the neural circuits. Training of the nervous system is ongoing; it happens with or without our conscious acknowledgement and regardless of its value for the individual: regularly used activity gets reinforced; unused gets forgotten. For instance, one can develop insomnia when sleep routine is regularly disrupted by shift work or chaotic lifestyle. However, reintroduction of sleep hygiene will assist restoration of normal night sleep patterns. The same principle is employed in biofeedback bowel [57] or bladder training [58] for certain ANS disorders. In this subsection we discuss a few important consequences of autonomic dysfunction which impact on chronic pain development.

Any medical condition is associated with disturbed function of one or another organ, and therefore, with the disturbance of autonomic regulation of the corresponding physiological system. The opposite statement is also true: disturbed autonomic regulation will cause symptom development (into a medical condition) or prevent recovery from a condition-inducing event. This could be applied to acute pain as a symptom of a condition in question, or to chronic pain as a disease on its own.

A degree of autonomic disturbance might vary with different types of pain, the part(s) of nervous system involved, and anatomical region(s) affected. ANS dysfunction can be of local or global significance, and of mild or more severe presentation. We can associate diseases limited by anatomical region with the corresponding typical pain picture, but systemic medical conditions (e.g., diabetes, cardiovascular and lung diseases, rheumatoid arthritis, sickle cell disease) contribute to many chronic pain states. That is why the situation with diagnosis of disease causation and with recognition of factors leading to pain chronicity is not straightforward. Traditionally, abnormal ANS function and its diagnosis is overlooked in many (especially somatic) pain states, therefore the prescribed treatment often addresses only local symptoms, rather than pathophysiology of the underlying mechanisms.

For many chronic diseases we should recognise a reversible preliminary phase with subtle signs, which are usually below the threshold of current medical tests. Over a period of time autonomic regulation becomes progressively abnormal, but due to built-in robustness of the ANS, clinically significant deviation from medical norms might manifest years after. The preliminary phase is not usually identified, and underlying issues are not corrected. Partially this is because subclinical signs do not fall into pathological zones, but rather into domains of fitness or risk factors. This is a field of preventive medicine which, unfortunately, is largely unfamiliar to the general public. The quality of life at this stage deteriorates slowly and patients usually adapt to these changes without noticing the ongoing problem.

The important question at this stage is whether a single organ autonomic dysfunction develops in isolation in a particular chronic pain state, or whether it is always a part of the more systemic trend. The diagnostic value of many available tests of ANS status in pre-clinical phase and in even in mild cases is questionable. Their results are frequently reported as negative (or mildly abnormal) as often subjective severity of symptoms of ANS dysfunction do not match objectively measured parameters [59].

From our clinical observations when patients are convinced that their symptoms fit into a picture of Postural Orthostatic Tachycardia Syndrome (POTS), interstitial cystitis or CRPS but investigations do not support their perceptions our attempts to reassure them often fail. At that point our misunderstanding of the situation, broken relationship with patient, and lack of tests with higher resolution or sensitivity (they define disease criteria) leads to delayed diagnosis and treatment. On the other hand, labelling patients with the above diagnoses without sufficient evidence may medicalise them for life and prevent recovery. This unfortunate dilemma is one of the innate weaknesses of medical practice. It is triggered by the patient's suffering from severe presenting symptoms.

The definition of suffering according to the Oxford English Dictionary is as follows: "the state of undergoing pain, distress, or hardship". In CP all three entities—pain, distress and hardship—are intertwined, making it difficult to address them. The relationship between chronic pain, distress and hardship is well recognised. It dwells in emotional, social, and behavioural domains, and often is maintained by general symptoms (fatigue, chest tightness, mental fog and memory disturbances, sleep disorders and many others).

Sometimes the above constellation of symptoms is explained as an affective component of pain (linking it to the use of the old somatic nociceptive pathways), which is only partially true, as in fully developed CP we deal with neuroplasticity of nervous system where the ANS is responsible for many of these consequences [60–62]. Supporting evidence from research shows sympathetic hyperactivity in mental fatigue [63], significant and substantial ANS role in memory consolidation during sleep [64], association of mental fog and autonomic hyperarousal [65], abnormal autonomic sleep regulation in CP [66], and activation of autonomic pathways for chest pain and dyspnoea [67].

When medical professionals meet distressed patients who do not have clearly visible pathology which could explain the high degree of suffering, they often refer to these cases as those with functional (neurological) symptoms. However, many of these 'unexplained' symptoms could be due to disorganised activity of the ANS. We support the idea that suffering in CP could not happen without inherent participation of the ANS. This is because of a few reasons. ANS reaction to any pain or insult is inseparable to pain, even if pain is of somatic origin. Chronicity of pain and suffering is always at least partially driven by local or global dysfunction of the ANS. This includes control by emotional and behavioural centres, and often is not related to the severity of the index trauma.

CP patients develop maladaptive emotions and demonstrate changed behaviour. This includes poor coping and passive [68] strategies, fear-avoidance, and lack of motivation to invest efforts for their recovery, and social withdrawal. The ANS plays an important role in these changes. An experimental study [69] demonstrated that visceral pain response might relate to personality type, and it discovered sympathetic and parasympathetic co-activation in response to somatic and visceral pain. It is well known that the longer the chronic pain condition lasts and the more prominent are the patient's passive approach, sick role, and other maladaptive psychological trends, the worse these features become, and patients with these symptoms are less likely to improve.

Finally, we should not forget that autonomic dysfunction in control of inflammation and immunity [33, 70–72], endocrine system [48], circadian rhythms [39, 73–77], tissue regeneration [78–80], including ANS itself [81] also contributes to chronic pain development and its chronicity.

Conclusion: The reciprocal relationship between pain and ANS control of involuntary body functions, emotions, behaviour and body regeneration makes the ANS an integral and indispensable player in a drama of CP. The earliest phase of autonomic dysfunction is not recognised and corrected.

6. Diagnostics

In this subsection we review diagnostic investigations currently available for assessment of ANS activity and discuss their limitations. We also describe potential tests (based on autonomic features) which might be applicable for pain assessment.

Heart rate variability (HRV) is one such non-invasive tool which is used in lab research, for diagnostic purposes, or in everyday life to monitor cardiorespiratory fitness. HRV employs electrocardiography (ECG) or plethysmography (PPG) for measuring distances between electrical heart complexes (or beats) over a period of time. The raw data obtained from ECG or PPG are calculated into various indexes which (as per convention) might describe activity of ANS branches.

Although HRV uses cardiac electrical activity for calculations, it shows not only good predictability of mortality in the heart conditions, but also demonstrates abnormalities in ANS performance in many diseases and pain states. However, HRV is not condition-specific. Additionally, there is no validated scale that can diagnose a degree of autonomic dysfunction in a particular illness.

HRV is a cheap, easy to use and widely accepted tool. For a full analysis it requires only a budget peripheral wearable device and a smartphone application. The analysis is based on mathematical calculations: descriptive statistics for time domain and spectral analysis for frequency domain. The latter uses the term "power" in relation to the energy within a particular frequency band, which should not be confused with biological "strength" of ANS divisions.

We do not know what the power of the ANS is and how to physically measure it. With HRV we might see a snapshot of the balance between sympathetic and parasympathetic activity. Whether this balance is on proportionally suppressed or enhanced divisions it is not possible to say. Furthermore, autonomic activity might be disturbed only in one organ or system, or in case of global autonomic failure, different organs might be affected unequally. These points should be taken into consideration when interpreting a HRV report in relation to a particular pathology.

There have been attempts to match HRV with organ-specific physiological activity. For instance, by parallel measurement of HRV and of high-resolution manometry of colon [82]. This experiment showed parasympathetic activation and sympathetic withdrawal during triggered propulsive colonic activity.

Clinical tests require laboratory conditions for measurements, calibrated and medically certified equipment, and professional interpretations. These tests investigate a single organ or a system specific autonomic dysfunction, but they are often invasive and might require anaesthetic input. For example, those used in cardiovascular medicine (e.g., tilt table test with plasma catecholamine concentration measurement), in urology (e.g., urodynamic tests), neurology (e.g., skin biopsy for nerve fibre density, nerve conduction studies, sudomotor activity and recordings of muscle sympathetic nerve activity), gastroenterology (e.g., gut motility, food transit, bacterial overgrowth).

Non-invasive options include disease specific questionnaires for organ function and thermography, a measurement of the surface temperature from the distance by thermal camera. The latter is useful in diseases with local change of blood supply, like in vascular abnormalities (vessel stenosis or arterio-venous malformation), regional sympathetic activity suppression (disease or local anaesthetic injection) or its excess (Raynaud's or iatrogenic). CP conditions which manifest with skin temperature changes include CRPS, neuropathic pain with neurogenic inflammation, ischaemic pain and some others [83].

Pupillometry (PPM) is another window into ANS activity. The size of the pupil depends upon the rhythmical activity of a sphincter (parasympathetic control) and dilator (sympathetic innervation), triggered by the amount of light reaching the retina. Despite a growing body of research in anaesthesia and acute perioperative pain management which use PPM for assessment of pain [84, 85] and drug effects, the utility of this non-invasive method has not been fully established.

Facial expressions (FE) and emotion recognition is a complex field (the ANS plays a major role in it) where stable prediction is not technically achieved. FE have been used in acute pain assessment for a long time, but not in CP. Computer vision techniques often employ facial action coding systems (FACS) which detect face geometry and movement patterns [86, 87].

We suggest that in CP sufferers, FE could be used to assess the effect of pain-relieving intervention. According to clinical observations (unpublished data of the first author—DK), successful interventions in cancer pain dramatically change the quality of FE. For example, if a patient smiles before a procedure, it looks unnatural, forced or laboured. When the pain is relieved by intervention the smile becomes more natural with genuine facial mimic. This is a promising area for research of the role of the ANS in CP and pain relief with an objective and quantitative outcome.

Parameters of voice and of speech change under stress, emotional and cognitive load, pathological conditions and via ANS influence [88, 89]. A few voice-forming and modulating muscles are innervated by autonomic motor nerve fibres. Voice analysis could be used for monitoring of therapy and prediction of deterioration during the course of disease [90]. It is becoming more popular for pain assessment with the arrival of Artificial Intelligence (AI) based software [91].

Conclusion: Testing ANS state is essential in CP management; it demonstrates universal autonomic participation in CP. Clinical tests could be condition-specific, but invasive and demanding (equipment, staff etc.). Non-invasive methods are becoming more available for personal use (HRV, thermography), but some are still under-developed (pupillometry, facial and voice analysis). The resolution of existing tests is still low for the early recognition of pathology. Tests give a cross-sectional view (snapshot) on the condition but cannot provide longitudinal data for the evaluation of underlying pathophysiology. The latter could be addressed with the use of wearable multi-modal biosensing systems [92, 93].

7. Treatment

This section bears a dual purpose. It speculates on how the ANS could shape the outcomes of conventional pain-relieving procedures, and highlights the potential therapeutic interventions for disturbed ANS control. We discuss peripheral nerve blocks, epidural blocks and neuromodulation.

Local anaesthetic (LA) of sufficient concentration blocks nerve conduction allowing painless surgery. Unfortunately, pain returns if nerve blockade fades away. However, for post-operative analgesia significantly lower concentrations of LA than for surgery are required as there is no ongoing tissue damage.

In CP an injection of Lidocaine (LA) could provide a relieving effect of significantly longer duration (sometimes for several months or years) than the length of the nerve block [94].

First, we would like to describe thoracic differential epidural (TDE) blockade which is used as a diagnostic tool for abdominal pain to discriminate between somatic, visceral or central pain, and to predict response to visceral nerve block [95]. This intervention exploits two facts: smaller diameter visceral nociceptive afferents are blocked by a lower concentration of LA; and pain relief in visceral pain lasts longer than anaesthetic block duration [96]. This intervention showed that in many patients with pancreatitis, pain is of somatic nature [97].

Using the example of TDE, we might generalise that the therapeutic effect of epidural or peripheral nerve injection (beyond the LA duration) for musculoskeletal (somatic) pain could be due to concomitant sympathetic blockade. Epidurals, nerve root injections and peripheral nerve blocks produce sympathetic blockade in corresponding dermatomes or nerve distributions [98, 99].

Spinal cord stimulation (SCS) can provide pain relief and improvement of other symptoms in visceral and somatic pain by neuromodulation of various targets within the spinal cord. For example, SCS for refractory abdominal pain can improve chronic nausea and vomiting [100]. For neuropathic visceral abdominal pain, clinicians target the upper-mid thoracic level where splanchnic nerves emerge from the spinal cord. Improvement of gastroparesis and intestinal motility is highly suggestive of sympathetic blockade produced by SCS. However, available studies do not demonstrate consistent ANS reaction to SCS in sudomotor activity [101], heart rate variability (HRV), baroreceptor reflex sensitivity (BRS) and muscle sympathetic nerve activity (MSNA). Nor do they provide a plausible hypothesis for mechanisms of pain relief related to autonomic control [102]. This is another important area for future research.

Similarly to SCS, sacral nerve stimulation (SaNS) for pelvic organ dysfunction (bladder and rectal control) can also result in improvement in pain control [103] related to treated conditions.

Percutaneous tibial (somatic peripheral nerve) nerve stimulation (PTNS), which is used to improve urinary bladder control in Overactive Bladder, is an effective and minimally invasive technique [104]. It requires multiple sessions to achieve prolonged effect. PTNS also relieves chronic pelvic pain [105]. The mechanism of action is unknown but clinically it improves autonomic reflexes of the targeted organs. We can speculate about two possibilities:

- Somatic afferent stimulation affects autonomic efferent output (this could be at the level of spinal cord, brain, or both);

- Antidromic stimulation of sympathetic fibres which supply skin and blood vessels produces this effect.

Percutaneous or surgical vagal nerve stimulation has been suggested for many conditions caused by or associated with autonomic dysfunction [106–112], but the main indications are refractory epilepsy and certain mood disorders.

Non-medical options to maintain healthy ANS activity, which frequently involves parasympathetic stimulation and sympathetic withdrawal, include:

- Meditation, controversial reports [113–115]
- Slow diaphragmatic breathing [116]
- Acclimation to cold exposure [117]
- Exercises [118–120]

Conclusion: When planning pain-relieving interventions, healthcare professionals should consider treatment of underlying ANS dysfunction.

8. Traditional medicine (TM)

Traditional medical practices always acknowledge the complex relationship between organs and somatic tissues. It is reflected in diagnostic methods and in the holistic approach to treatment.

One of the unique methods used by many systems is the pulse diagnostic tool. It requires years to master but is claimed to provide invaluable information about any organ in the body. In general, it is probably an ancient equivalent of HRV, but much more sophisticated in the amount of detailed information it might provide an experienced practitioner.

Many traditions use a quasi-anatomical system: a whole-body map covered with lines or meridians (channels or vessels) with named points with very precise locations. The 'vital energy' freely flows through these structures controlling the activity of different physiological systems (lungs, bowel, liver, stomach etc.). There are 12 principal meridians in Traditional Chinese Medicine (TCM). There is no equivalent concept in western medicine.

The 24-hour biorhythm cycle allocates time of the highest and of the lowest activity to each channel. For example, the first meridian (Lungs) is the most active between 3 and 5 AM and 12 hours later (3 and 5 PM) it is in the lowest energy state. This circadian clock schedule correlates with clinical observations. For instance, maximum activity in the lung meridian corresponds to the peak of nocturnal asthma attacks. Maximum activity of the second meridian (Bowel) falls between 5 and 7 AM when people wake up after the night sleep and open their bowel. The next meridian (Stomach) is the most active when people normally have their breakfast, between 7 and 9 AM, and so on.

Abnormal flow of 'energy' (deficiency or excess and blockage) in one or more channels is the reason for symptoms of disease. Needling of the points according to the acupuncture recipe restores 'energy' flow and cures the disease and relieves pain (Yuan 2015). The choice of acupoints depends on the diagnosis, biorhythms and relationship between meridians. For example, neck and shoulder pain could be related to abnormal energy situations in gallbladder, bladder or large intestine meridians. Points could be chosen on these meridians or on others (via laws of relationship), around the painful area or distant to it; and they vary on a different time or day.

There are, of course, other than acupuncture treatment methods in TM: medications, breathing practices, postures and movements. The latter two could be organ specific, and they are synchronised with breathing to optimise vital energy and its circulation.

Conclusion: TM uses a holistic approach. It operates at the levels of aetiology and pathophysiology rather than symptoms as western medicine does. Similarly to the ANS, the Meridian system functions according to biological rhythms and connects organs and somatic tissues (skin, muscles, bones, ligaments and joints).

9. Conclusion

CP is considered incurable as per current beliefs, personal experience of medical professionals and statistics. So, in a pain clinic, in the media and in professional literature, patients once diagnosed with CP receive the same message, they have to live with it. The present situation is maintained by the ignorance of already known facts. However, progress is being fuelled by breakthroughs in related fields across multiple disciplines. We propose that pain becomes chronic through significant input from the ANS. Autonomic dysfunction (subclinical or apparent, local or global) provides the background for suboptimal organ activity and subsequently leads to the development of chronic symptoms, including pain, or the transition from existing acute pain into CP.

We would like to bring attention to the mind-blowing complexity and ample presence [48, 49, 121] of the ANS in our lives, as well as to its important role in CP and pain chronicity. This chapter serves only to outline a topic which could easily fill a whole book and warrants ongoing research. Current evidence from evolutionary biology, anatomy, epidemiology, pathophysiology, diagnostics, and pain medicine (western and traditional) supports our view and paves the way for future work. However, even now, people might change their view on the topic and this could lead to improved outcomes in CP management. Therefore, the main takeaway message is that we have to seek the signs of ANS dysfunction in any CP condition and address the underlying mechanisms.

Conflict of interest

The authors declare no conflict of interest.

Author details

Dmitry Kruglov[1,2]* and Dermot McGuckin[1,3]

1 Pain Management Centre, National Hospital for Neurology and Neurosurgery, London, UK

2 University College London Hospital, London, UK

3 Research Department of Targeted Intervention, Division of Surgery and Interventional Science, University College London, London, UK

*Address all correspondence to: dg_kruglov@hotmail.com

References

[1] Fayaz A, Croft P, Langford RM, Donaldson LJ, Jones GT. Prevalence of chronic pain in the UK: A systematic review and meta-analysis of population studies. BMJ Open. 2016;**6**(6):e010364. DOI: 10.1136/bmjopen-2015-010364

[2] Sá KN, Moreira L, Baptista AF, Yeng LT, Teixeira MJ, Galhardoni R, et al. Prevalence of chronic pain in developing countries: Systematic review and meta-analysis. Pain Reports. 2019;**4**(6):e779. DOI: 10.1097/PR9.0000000000000779

[3] Mills SEE, Nicolson KP, Smith BH. Chronic pain: a review of its epidemiology and associated factors in population-based studies. British Journal of Anaesthesia. 2019;**123**(2):e273-e283. DOI: 10.1016/j.bja.2019.03.023

[4] McGuire DB. Comprehensive and multidimensional assessment and measurement of pain. Journal of Pain and Symptom Management. 1992;7(5):312-319. DOI: 10.1016/0885-3924(92)90064-o

[5] Kyle BN, McNeil DW. Autonomic arousal and experimentally induced pain: A critical review of the literature. Pain Research & Management. 2014;**19**(3):159-167. DOI: 10.1155/2014/536859

[6] Arendt D, Denes AS, Jékely G, Tessmar-Raible K. The evolution of nervous system centralization. Philosophical Transactions of the Royal Society of London. Series B, Biological Sciences. 2008;**363**(1496):1523-1528. DOI: 10.1098/rstb.2007.2242

[7] Northcutt RG. Evolution of centralized nervous systems: two schools of evolutionary thought. Proceedings of the National Academy of Sciences of the United States of America. 2012;**109**(Suppl 1):10626-10633. DOI: 10.1073/pnas.1201889109

[8] Porges SW, Furman SA. The early development of the autonomic nervous system provides a neural platform for social behavior: A polyvagal perspective. Infant and Child Development. 2011;**20**(1):106-118. DOI: 10.1002/icd.688

[9] Fritzsch B, Elliott KL, Glover JC. Gaskell revisited: New insights into spinal autonomics necessitate a revised motor neuron nomenclature. Cell and Tissue Research. 2017;**370**(2):195-209. DOI: 10.1007/s00441-017-2676-y

[10] Wang T. Chapter 141—Evolution of the cardiovascular autonomic nervous system in vertebrates. In: Robertson D, Biaggioni I, Burnstock G, Low PA, Paton JFR, editors. Primer on the Autonomic Nervous System. Third ed. Academic Press; 2012. pp. 669-673. DOI: 10.1016/B978-0-12-386525-0.00141-4

[11] Nomaksteinsky M, Kassabov S, Chettouh Z, Stoeklé HC, Bonnaud L, Fortin G, et al. Ancient origin of somatic and visceral neurons. BMC Biology. 2013;**11**:53. DOI: 10.1186/1741-7007-11-53

[12] Jänig W, Häbler HJ. Neurophysiological analysis of target-related sympathetic pathways—From animal to human: Similarities and differences. Acta Physiologica Scandinavica. 2003;**177**(3):255-274. DOI: 10.1046/j.1365-201X.2003.01088.x

[13] Jänig W, Häbler HJ. Specificity in the organization of the autonomic nervous system: A basis for precise neural regulation of homeostatic and protective body functions. Progress

in Brain Research. 2000;**122**:351-367. DOI: 10.1016/s0079-6123(08)62150-0

[14] Jänig W. Neurobiology of visceral afferent neurons: Neuroanatomy, functions, organ regulations and sensations. Biological Psychology. 1996;**42**(1-2):29-51. DOI: 10.1016/0301-0511(95)05145-7

[15] Dauger S, Pattyn A, Lofaso F, Gaultier C, Goridis C, Gallego J, et al. Phox2b controls the development of peripheral chemoreceptors and afferent visceral pathways. Development. 2003;**130**(26):6635-6642. DOI: 10.1242/dev.00866

[16] Bertucci P, Arendt D. Somatic and visceral nervous systems—An ancient duality. BMC Biology. 2013;**11**:54. DOI: 10.1186/1741-7007-11-54

[17] Mazzone SB, Undem BJ. Vagal afferent innervation of the airways in health and disease. Physiological Reviews. 2016;**96**(3):975-1024. DOI: 10.1152/physrev.00039.2015

[18] Kupari J, Häring M, Agirre E, Castelo-Branco G, Ernfors P. An atlas of vagal sensory neurons and their molecular specialization. Cell Reports. 2019;**27**(8):2508-2523.e4. DOI: 10.1016/j.celrep.2019.04.096

[19] Wang YB, de Lartigue G, Page AJ. Dissecting the role of subtypes of gastrointestinal vagal afferents. Frontiers in Physiology. 2020;**11**:643. DOI: 10.3389/fphys.2020.00643

[20] Li X, Wang L. Rethinking the visceral innervation—Peek into the emerging field of molecular dissection of neural signals. Biochemical and Biophysical Research Communications. 2022;**633**:20-22. DOI: 10.1016/j.bbrc.2022.09.011

[21] Jänig W. Role of visceral afferent neurons in visceral nociception and pain. In: Integrative Action of the Autonomic Nervous System: Neurobiology of Homeostasis. Cambridge: Cambridge University Press; 2006. pp. 54-60. DOI: 10.1017/CBO9780511541667.007

[22] Basbaum A. History of spinal cord "Pain" pathways including the pathways not taken. Frontiers in Pain Research (Lausanne). 2022;**3**:910954. DOI: 10.3389/fpain.2022.910954

[23] Westlund KN. Visceral nociception. Current Review of Pain. 2000;**4**(6):478-487. DOI: 10.1007/s11916-000-0072-9

[24] Houghton AK, Wang CC, Westlund KN. Do nociceptive signals from the pancreas travel in the dorsal column? Pain. 2001;**89**(2-3):207-220. DOI: 10.1016/s0304-3959(00)00364-x

[25] Arslan D, Ünal ÇI. Interactions between the painful disorders and the autonomic nervous system. Agri. 2022;**34**(3):155-165. DOI: 10.14744/agri.2021.43078

[26] Fayaz A, Ayis S, Panesar SS, Langford RM, Donaldson LJ. Assessing the relationship between chronic pain and cardiovascular disease: A systematic review and meta-analysis. Scandinavian Journal of Pain. 2016;**13**:76-90. DOI: 10.1016/j.sjpain.2016.06.005

[27] Shen Q, Guo T, Song M, et al. Pain is a common problem in patients with ILD. Respiratory Research. 2020;**21**:297. DOI: 10.1186/s12931-020-01564-0

[28] Nair SP, Panchabhai CS, Panhale V. Chronic neck pain and respiratory dysfunction: A review paper. Bulletin of Faculty of Physical Therapy. 2022;**27**:21. DOI: 10.1186/s43161-022-00078-8

[29] Falling C, Stebbings S, Baxter GD, Gearry RB, Mani R. Musculoskeletal pain in individuals with inflammatory bowel disease reflects three distinct profiles. The Clinical Journal of Pain. 2019;**35**(7):559-568. DOI: 10.1097/AJP.0000000000000698

[30] Charrua A, Pinto R, Birder LA, Cruz F. Sympathetic nervous system and chronic bladder pain: A new tune for an old song. Translational Andrology and Urology. 2015;**4**(5):534-542. DOI: 10.3978/j.issn.2223-4683.2015.09.06

[31] Flegge LG, Barr A, Craner JR. Sexual functioning among adults with chronic pain: Prevalence and association with pain-related outcomes. Pain Medicine. 2023;**24**(2):197-206. DOI: 10.1093/pm/pnac117

[32] Rittner HL, Oehler B, Stein C. 5.23—Immune system, pain and analgesia. In: Fritzsch B, editor. The Senses: A Comprehensive Reference. Second ed. Elsevier; 2020. pp. 385-397. DOI: 10.1016/B978-0-12-809324-5.24129-1

[33] Larson AA, Pardo JV, Pasley JD. Review of overlap between thermoregulation and pain modulation in fibromyalgia. The Clinical Journal of Pain. 2014;**30**(6):544-555. DOI: 10.1097/AJP.0b013e3182a0e383

[34] Jänig W. Relationship between pain and autonomic phenomena in headache and other pain conditions. Cephalalgia. 2003;**23**(Suppl 1):43-48. DOI: 10.1046/j.1468-2982.2003.00573.x

[35] Sundström T, Guez M, Hildingsson C, Toolanen G, Nyberg L, Riklund K. Altered cerebral blood flow in chronic neck pain patients but not in whiplash patients: a 99mTc-HMPAO rCBF study. European Spine Journal. 2006;**15**(8):1189-1195. DOI: 10.1007/s00586-005-0040-5

[36] Iwabuchi SJ, Xing Y, Cottam WJ, Drabek MM, Tadjibaev A, Fernandes GS, et al. Brain perfusion patterns are altered in chronic knee pain: A spatial covariance analysis of arterial spin labelling MRI. Pain. 2020;**161**(6):1255-1263. DOI: 10.1097/j.pain.0000000000001829

[37] de Zambotti M, Baker FC. Sleep and circadian regulation of the autonomic nervous system. In: Autonomic nervous system and sleep. Cham: Springer; 2021. p.63-69. DOI: 10.1007/978-3-030-62263-3

[38] Bumgarner JR, Walker WH, Nelson RJ. Circadian rhythms and pain. Neuroscience and Biobehavioral Reviews. 2021;**129**:296-306. DOI: 10.1016/j.neubiorev.2021.08.004

[39] Seymour B. Pain: A precision signal for reinforcement learning and control. Neuron. 2019;**101**(6):1029-1041. DOI: 10.1016/j.neuron.2019.01.055

[40] Horing B, Büchel C. The human insula processes both modality-independent and pain-selective learning signals. PLoS Biology. 2022;**20**(5):e3001540. DOI: 10.1371/journal.pbio.3001540

[41] Stanisic N, Häggman-Henrikson B, Kothari M, Costa YM, Avivi-Arber L, Svensson P. Pain's adverse impact on training-induced performance and neuroplasticity: A systematic review. Brain Imaging and Behavior. 2022;**16**(5):2281-2306. DOI: 10.1007/s11682-021-00621-6

[42] Matthews D, Cancino EE, Falla D, Khatibi A. Exploring pain interference with motor skill learning in humans: A systematic review. PLoS One. 2022;**17**(9):e0274403. DOI: 10.1371/journal.pone.0274403

[43] Fine PG. Long-term consequences of chronic pain: Mounting evidence for pain

as a neurological disease and parallels with other chronic disease states. Pain Medicine. 2011;**12**(7):996-1004. DOI: 10.1111/j.1526-4637.2011.01187.x

[44] Glatte P, Buchmann SJ, Hijazi MM, Illigens BM, Siepmann T. Architecture of the cutaneous autonomic nervous system. Frontiers in Neurology. 2019;**10**:970. DOI: 10.3389/fneur.2019.00970

[45] Khan MM, Lustrino D, Silveira WA, Wild F, Straka T, Issop Y, et al. Sympathetic innervation controls homeostasis of neuromuscular junctions in health and disease. Proceedings of the National Academy of Sciences of the United States of America. 2016;**113**(3):746-750. DOI: 10.1073/pnas.1524272113

[46] Rodrigues ACZ, Messi ML, Wang ZM, Abba MC, Pereyra A, Birbrair A, et al. The sympathetic nervous system regulates skeletal muscle motor innervation and acetylcholine receptor stability. Acta Physiologica (Oxford, England). 2019;**225**(3):e13195. DOI: 10.1111/apha.13195

[47] Jänig W. 5.21 - Sympathetic Nervous System and Pain. In: Fritzsch B, Editor-in-Chief. The Senses: A Comprehensive Reference. 2nd ed. London: Elsevier Inc.; 2021. p.349-378. DOI: 10.1016/B978-0-12-809324-5.24248-X

[48] Jänig W. The peripheral sympathetic and parasympathetic pathways. In: Integrative Action of the Autonomic Nervous System: Neurobiology of Homeostasis. Cambridge: Cambridge University Press; 2006. pp. 106-167. DOI: 10.1017/CBO9780511541667.007

[49] Fleckenstein J, Neuberger EWI, Bormuth P, Comes F, Schneider A, Banzer W, et al. Investigation of the sympathetic regulation in delayed onset muscle soreness: Results of an RCT. Frontiers in Physiology. 2021;**12**:697335. DOI: 10.3389/fphys.2021.697335

[50] Egli S, Pfister M, Ludin SM, Puente de la Vega K, Busato A, Fischer L. Long-term results of therapeutic local anesthesia (neural therapy) in 280 referred refractory chronic pain patients. BMC Complementary and Alternative Medicine. 2015;**15**:200. DOI: 10.1186/s12906-015-0735-z

[51] Fischer L, Barop H, Ludin SM, Schaible HG. Regulation of acute reflectory hyperinflammation in viral and other diseases by means of stellate ganglion block. A conceptual view with a focus on Covid-19. Autonomic Neuroscience. 2022;**237**:102903. DOI: 10.1016/j.autneu.2021.102903

[52] Bartley EJ, Fillingim RB. Sex differences in pain: A brief review of clinical and experimental findings. British Journal of Anaesthesia. 2013;**111**(1):52-58. DOI: 10.1093/bja/aet127

[53] Dart AM, Du XJ, Kingwell BA. Gender, sex hormones and autonomic nervous control of the cardiovascular system. Cardiovascular Research. 2002;**53**(3):678-687. DOI: 10.1016/s0008-6363(01)00508-9

[54] Zhang W, Liu Y, Xu J, Fan C, Zhang B, Feng P, et al. The role of sympathetic nerves in osteoporosis: A narrative review. Biomedicine. 2022;**11**(1):33. DOI: 10.3390/biomedicines11010033

[55] Christensen JS, Wild H, Kenzie ES, Wakeland W, Budding D, Lillas C. Diverse autonomic nervous system stress response patterns in childhood sensory modulation. Frontiers in Integrative Neuroscience. 2020;**14**:6. DOI: 10.3389/fnint.2020.00006

[56] Rao SS. Biofeedback therapy for constipation in adults. Best Practice & Research. Clinical Gastroenterology. 2011;**25**(1):159-166. DOI: 10.1016/j.bpg.2011.01.004

[57] Wagner B, Steiner M, Huber DFX, et al. The effect of biofeedback interventions on pain, overall symptoms, quality of life and physiological parameters in patients with pelvic pain. Wiener Klinische Wochenschrift. 2022;**134**(Suppl 1):11-48. DOI: 10.1007/s00508-021-01827-w

[58] Vincent A, Whipple MO, Low PA, Joyner M, Hoskin TL. Patients with fibromyalgia have significant autonomic symptoms but modest autonomic dysfunction. PM & R : The Journal of Injury, Function, and Rehabilitation. 2016;**8**(5):425-435. DOI: 10.1016/j.pmrj.2015.08.008

[59] Okamoto LE, Raj SR, Peltier A, Gamboa A, Shibao C, Diedrich A, et al. Neurohumoral and haemodynamic profile in postural tachycardia and chronic fatigue syndromes. Clinical Science (London, England). 2012;**122**(4):183-192. DOI: 10.1042/CS20110200

[60] Tanaka M, Tajima S, Mizuno K, Ishii A, Konishi Y, Miike T, et al. Frontier studies on fatigue, autonomic nerve dysfunction, and sleep-rhythm disorder. The Journal of Physiological Sciences. 2015;**65**(6):483-498. DOI: 10.1007/s12576-015-0399-y

[61] Wu REY, Khan FM, Hockin BCD, Lobban TCA, Sanatani S, Claydon VE. Faintly tired: A systematic review of fatigue in patients with orthostatic syncope. Clinical Autonomic Research. 2022;**32**(3):185-203. DOI: 10.1007/s10286-022-00868-z

[62] Mizuno K, Tanaka M, Yamaguti K, Kajimoto O, Kuratsune H, Watanabe Y. Mental fatigue caused by prolonged cognitive load associated with sympathetic hyperactivity. Behavioral and Brain Functions. 2011;7:17. DOI: 10.1186/1744-9081-7-17

[63] Whitehurst LN et al. Autonomic activity during sleep predicts memory consolidation in humans. Proceedings of the National Academy of Sciences of the United States of America. 2016;**113**(26):7272-7277. DOI: 10.1073/pnas.1518202113

[64] Rodriguez B, Hochstrasser A, Eugster PJ, Grouzmann E, Müri RM, Z'Graggen WJ. Brain fog in neuropathic postural tachycardia syndrome may be associated with autonomic hyperarousal and improves after water drinking. Frontiers in Neuroscience. 2022;**16**:968725. DOI: 10.3389/fnins.2022.968725

[65] Lavigne G, Okura K, Smith M. 5.20 - Pain Perception – Nociception During Sleep. In: Fritzsch B, Editor-in-Chief. The Senses: A Comprehensive Reference. 2nd ed. London: Elsevier Inc.; 2021. p.340-348. DOI: 10.1016/B978-0-12-805408-6.00195-0

[66] Burki NK, Lee LY. Mechanisms of dyspnea. Chest. 2010;**138**(5):1196-1201. DOI: 10.1378/chest.10-0534

[67] Bandler R, Keay KA, Floyd N, Price J. Central circuits mediating patterned autonomic activity during active vs. passive emotional coping. Brain Research Bulletin. 2000;**53**(1):95-104. DOI: 10.1016/s0361-9230(00)00313-0

[68] Paine P, Kishor J, Worthen SF, Gregory LJ, Aziz Q. Exploring relationships for visceral and somatic pain with autonomic control and personality. Pain. 2009;**144**(3):236-244. DOI: 10.1016/j.pain.2009.02.022

[69] Wei Y, Liang Y, Lin H, et al. Autonomic nervous system and

inflammation interaction in endometriosis-associated pain. Journal of Neuroinflammation. 2020;**17**:80. DOI: 10.1186/s12974-020-01752-1

[70] Bellocchi C, Carandina A, Montinaro B, Targetti E, Furlan L, Rodrigues GD, et al. The interplay between autonomic nervous system and inflammation across systemic autoimmune diseases. International Journal of Molecular Sciences. 2022;**23**(5):2449. DOI: 10.3390/ijms23052449

[71] Wasker SVZ, Challoumas D, Weng W, Murrell GAC, Millar NL. Is neurogenic inflammation involved in tendinopathy? A systematic review. BMJ Open Sport & Exercise Medicine. 2023;**9**(1):e001494. DOI: 10.1136/bmjsem-2022-001494

[72] Riganello F, Prada V, Soddu A, di Perri C, Sannita WG. Circadian rhythms and measures of CNS/autonomic interaction. International Journal of Environmental Research and Public Health. 2019;**16**(13):2336. DOI: 10.3390/ijerph16132336

[73] Warfield AE, Prather JF, Todd WD. Systems and circuits linking chronic pain and circadian rhythms. Frontiers in Neuroscience. 2021;**15**:705173. DOI: 10.3389/fnins.2021.705173

[74] Navarro-Ledesma S, Gonzalez-Muñoz A, García Ríos MC, de la Serna D, Pruimboom L. Circadian variation of blood pressure in patients with chronic musculoskeletal pain: A cross-sectional study. International Journal of Environmental Research and Public Health. 2022;**19**(11):6481. DOI: 10.3390/ijerph19116481

[75] Knezevic NN, Nader A, Pirvulescu I, Pynadath A, Rahavard BB, Candido KD. Circadian pain patterns in human pain conditions—A systematic review. Pain Practice. 2023;**23**(1):94-109. DOI: 10.1111/papr.13149

[76] Bumgarner JR, McCray EW, Nelson RJ. The disruptive relationship among circadian rhythms, pain, and opioids. Frontiers in Neuroscience. 2023;**17**:1109480. DOI: 10.3389/fnins.2023.1109480

[77] Knox SM, Lombaert IM, Haddox CL, Abrams SR, Cotrim A, Wilson AJ, et al. Parasympathetic stimulation improves epithelial organ regeneration. Nature Communications. 2013;**4**:1494. DOI: 10.1038/ncomms2493

[78] Davis EA, Dailey MJ. A direct effect of the autonomic nervous system on somatic stem cell proliferation? American Journal of Physiology. Regulatory, Integrative and Comparative Physiology. 2019;**316**(1):R1-R5. DOI: 10.1152/ajpregu.00266.2018

[79] Zhang Z, Hao Z, Xian C, Fang Y, Cheng B, Wu J, et al. Neuro-bone tissue engineering: Multiple potential translational strategies between nerve and bone. Acta Biomaterialia. 2022;**153**:1-12. DOI: 10.1016/j.actbio.2022.09.023

[80] Muratori L, Fregnan F, Carta G, Geuna S. Autonomic nervous system repair and regeneration. In: Phillips J, Hercher D, Hausner T, editors. Peripheral Nerve Tissue Engineering and Regeneration. Reference Series in Biomedical Engineering(). Cham: Springer; 2021. DOI: 10.1007/978-3-030-06217-0_2-1

[81] Ali MK, Liu L, Chen JH, Huizinga JD. Optimizing autonomic function analysis via heart rate variability associated with motor activity of the human colon. Frontiers in Physiology. 2021;**12**:619722. DOI: 10.3389/fphys.2021.619722

[82] Nahm FS. Infrared thermography in pain medicine. The Korean Journal of Pain. 2013;**26**(3):219-222. DOI: 10.3344/kjp.2013.26.3.219

[83] Thomsen LL, Olesen J. Autonomic aspects of migraine: Pathophysiology and treatment. In: Mathias CJ, Bannister SR, editors. Autonomic Failure: A Textbook of Clinical Disorders of the Autonomic Nervous System. 5th ed. Oxford: Oxford Academic; 2013, 2013. DOI: 10.1093/med/9780198566342.003.0070

[84] Wildemeersch D, Peeters N, Saldien V, Vercauteren M, Hans G. Pain assessment by pupil dilation reflex in response to noxious stimulation in anaesthetized adults. Acta Anaesthesiologica Scandinavica. 2018;**62**(8):1050-1056. DOI: 10.1111/aas.13129

[85] Prkachin KM. Assessing pain by facial expression: Facial expression as nexus. Pain Research & Management. 2009;**14**(1):53-58. DOI: 10.1155/2009/542964

[86] Naranjo-Hernández D, Reina-Tosina J, Roa LM. Sensor technologies to manage the physiological traits of chronic pain: A review. Sensors (Basel). 2020;**20**(2):365. DOI: 10.3390/s20020365

[87] MacPherson MK, Abur D, Stepp CE. Acoustic measures of voice and physiologic measures of autonomic arousal during speech as a function of cognitive load. Journal of Voice. 2017;**31**(4):504.e1-504.e9. DOI: 10.1016/j.jvoice.2016.10.021

[88] Van Puyvelde M, Neyt X, McGlone F, Pattyn N. Voice stress analysis: A new framework for voice and effort in human performance. Frontiers in Psychology. 2018;**9**:1994. DOI: 10.3389/fpsyg.2018.01994

[89] Sara JDS, Maor E, Borlaug B, Lewis BR, Orbelo D, et al. Non-invasive vocal biomarker is associated with pulmonary hypertension. PLoS One. 2020;**15**(4):e0231441. DOI: 10.1371/journal.pone.0231441

[90] Borna S, Haider CR, Maita KC, Torres RA, Avila FR, Garcia JP, et al. A review of voice-based pain detection in adults using artificial intelligence. Bioengineering (Basel). 2023;**10**(4):500. DOI: 10.3390/bioengineering10040500

[91] Siddharth PAN, Jung TP, Sejnowski TJ. A wearable multi-modal bio-sensing system towards real-world applications. IEEE Transactions on Biomedical Engineering. 2019;**66**(4):1137-1147. DOI: 10.1109/TBME.2018.2868759

[92] Leroux A, Rzasa-Lynn R, Crainiceanu C, Sharma T. Wearable devices: Current status and opportunities in pain assessment and management. Digit Biomark. 2021;**5**(1):89-102. DOI: 10.1159/000515576

[93] John NB, Hodgden J. Epidural injections for long term pain relief in lumbar spinal stenosis. The Journal of the Oklahoma State Medical Association. 2019;**112**(6):158-159

[94] Rizk MK, Tolba R, Kapural L, Mitchell J, Lopez R, Mahboobi R, et al. Differential epidural block predicts the success of visceral block in patients with chronic visceral abdominal pain. Pain Practice. 2012;**12**(8):595-601. DOI: 10.1111/j.1533-2500.2012.00548.x

[95] Jaffe RA, Rowe MA. Differential nerve block: Direct measurements on individual myelinated and unmyelinated dorsal root axons. Anesthesiology. 1996;**84**:1455-1464. DOI: 10.1097/00000542-199606000-00022

[96] Conwell DL, Vargo JJ, Zuccaro G, Dews TE, Mekhail N, Scheman J, et al. Role of differential neuroaxial blockade in the evaluation and management of pain in chronic pancreatitis. The American Journal of Gastroenterology. 2001;**96**(2):431-436. DOI: 10.1111/j.1572-0241.2001.03459.x

[97] Lange KH, Jansen T, Asghar S, Kristensen PL, Skjønnemand M, Nørgaard P. Skin temperature measured by infrared thermography after specific ultrasound-guided blocking of the musculocutaneous, radial, ulnar, and median nerves in the upper extremity. British Journal of Anaesthesia. 2011;**106**(6):887-895. DOI: 10.1093/bja/aer085

[98] Kruglov D, Stricker R, Howell K. Study of pattern of feet skin temperature distribution during continuous post-operative epidural analgesia. Proceedings of the 2020 International Conference on Quantitative InfraRed Thermography. 2023. DOI: 10.21611/qirt.2020.060. Available from: http://www.qirt.org/dynamique/index.php?idD=88&Lang=0

[99] Kapural L, Brown BK, Harandi S, Rejeski J, Koch K. Effects of spinal cord stimulation in patients with chronic nausea, vomiting, and refractory abdominal pain. Digestive Diseases and Sciences. 2022;**67**(2):598-605. DOI: 10.1007/s10620-021-06896-5

[100] Goudman L, Vets N, Jansen J, De Smedt A, Billot M, Rigoard P, et al. Electrochemical skin conductance alterations during spinal cord stimulation: An experimental study. Journal of Clinical Medicine. 2021;**10**(16):3565. DOI: 10.3390/jcm10163565

[101] Black S, Bretherton B, Baranidharan B, Murray A, Crowther T, Deuchars S, et al. A feasibility study exploring measures of autonomic function in patients with failed Back surgery syndrome undergoing spinal cord stimulation. Neuromodulation: Technology at the Neural Interface. 2023;**26**(1):192-205. DOI: 10.1016/j.neurom.2021.10.016

[102] Greig J, Mak Q, Furrer MA, Sahai A, Raison N. Sacral neuromodulation in the management of chronic pelvic pain: A systematic review and meta-analysis. Neurourology and Urodynamics. 2023;**42**(4):822-836. DOI: 10.1002/nau.25167

[103] Gaziev G, Topazio L, Iacovelli V, Asimakopoulos A, Di Santo A, De Nunzio C, et al. Percutaneous Tibial Nerve Stimulation (PTNS) efficacy in the treatment of lower urinary tract dysfunctions: A systematic review. BMC Urology. 2013;**13**:61. DOI: 10.1186/1471-2490-13-61

[104] Sevim M, Alkiş O, Kartal İG, Kazan HO, İvelik Hİ, Aras B, et al. Comparison of transcutaneous tibial nerve stimulation versus percutaneous tibial nerve stimulation in category IIIB chronic prostatitis/chronic pelvic pain syndrome: A randomized prospective trial. The Prostate. 2023;**83**(8):751-758. DOI: 10.1002/pros.24513

[105] Bonaz B, Sinniger V, Pellissier S. The vagus nerve in the neuro-immune axis: Implications in the pathology of the gastrointestinal tract. Frontiers in Immunology. 2017;**8**:1452. DOI: 10.3389/fimmu.2017.01452

[106] Chang RB. Body thermal responses and the vagus nerve. Neuroscience Letters. 2019;**698**:209-216. DOI: 10.1016/j.neulet.2019.01.013

[107] Chang YC, Ahmed U, Jayaprakash N, Mughrabi I, Lin Q, Wu YC, et al. kHz-frequency electrical stimulation selectively activates small, unmyelinated vagus afferents. Brain Stimulation. 2022;**15**(6):1389-1404. DOI: 10.1016/j.brs.2022.09.015

[108] Imai J, Katagiri H. Regulation of systemic metabolism by the autonomic nervous system consisting of afferent and efferent innervation. International Immunology. 2022;**34**(2):67-79. DOI: 10.1093/intimm/dxab023

[109] Ahmed U, Chang YC, Zafeiropoulos S, Nassrallah Z, Miller L, Zanos S. Strategies for precision vagus neuromodulation. Bioelectronic Medicine. 2022;**8**(1):9. DOI: 10.1186/s42234-022-00091-1

[110] Falvey A. Vagus nerve stimulation and inflammation: Expanding the scope beyond cytokines. Bioelectronic Medicine. 2022;**8**(1):19. DOI: 10.1186/s42234-022-00100-3

[111] Farrand A, Jacquemet V, Verner R, Owens M, Beaumont E. Vagus nerve stimulation parameters evoke differential neuronal responses in the locus coeruleus. Physiological Reports. 2023;**11**(5):e15633. DOI: 10.14814/phy2.15633

[112] Adler-Neal AL, Waugh CE, Garland EL, Shaltout HA, Diz DI, Zeidan F. The role of heart rate variability in mindfulness-based pain relief. The Journal of Pain. 2020;**21**(3-4):306-323. DOI: 10.1016/j.jpain.2019.07.003

[113] Ganguly A, Hulke SM, Bharshanakar R, Parashar R, Wakode S. Effectofmeditationonautonomicfunction in healthy individuals: A longitudinal study. Journal of Family Medicine and Primary Care. 2020;**9**(8):3944-3948. DOI: 10.4103/jfmpc.jfmpc_460_20

[114] Natarajan A. Heart rate variability during mindful breathing meditation. Frontiers in Physiology. 2023;**13**:1017350. DOI: 10.3389/fphys.2022.1017350

[115] Busch V, Magerl W, Kern U, Haas J, Hajak G, Eichhammer P. The effect of deep and slow breathing on pain perception, autonomic activity, and mood processing—An experimental study. Pain Medicine. 2012;**13**(2):215-228. DOI: 10.1111/j.1526-4637.2011.01243.x

[116] Mäkinen TM, Mäntysaari M, Pääkkönen T, Jokelainen J, Palinkas LA, Hassi J, et al. Autonomic nervous function during whole-body cold exposure before and after cold acclimation. Aviation, Space, and Environmental Medicine. 2008;**79**(9):875-882. DOI: 10.3357/asem.2235.2008

[117] Araujo CG, Laukkanen JA. Heart and skeletal muscles: Linked by autonomic nervous system. Arquivos Brasileiros de Cardiologia. 2019;**112**(6):747-748. DOI: 10.5935/abc.20190097

[118] Abuín-Porras V, Clemente-Suárez VJ, Jaén-Crespo G, Navarro-Flores E, Pareja-Galeano H, Romero-Morales C. Effect of physiotherapy treatment in the autonomic activation and pain perception in male patients with non-specific subacute low back pain. Journal of Clinical Medicine. 2021;**10**(8):1793. DOI: 10.3390/jcm10081793

[119] Daniela M, Catalina L, Ilie O, Paula M, Daniel-Andrei I, Ioana B. Effects of exercise training on the autonomic nervous system with a focus on anti-inflammatory and antioxidants effects. Antioxidants (Basel). 2022;**11**(2):350. DOI: 10.3390/antiox11020350

[120] Yuan QL, Guo TM, Liu L, Sun F, Zhang YG. Traditional Chinese medicine for neck pain and low back pain: A systematic review and meta-analysis. PLoS One. 2015;**10**(2):e0117146. DOI: 10.1371/journal.pone.0117146

[121] Brodal P. The Central Nervous System. 5th ed. New York: Oxford Academic; 2016, 2016. DOI: 10.1093/med/9780190228958.001.0001

Chapter 4

Exploring Cardiac Responses of Pain and Distress

Mona Elsayed and Elizabeth Barbara Torres

Abstract

Pain and distress stand at the intersection of multiple health crises and are leading contributors to disability. Current pain assessments rely on self-reports—which assume a capacity to understand and verbalize mental/emotional states—and behavioral observation which can be subject to limitations and misinterpretation. Methods to evaluate pain/distress can be substantially enhanced with biometrics that incorporate the physiological aspects of the full pain experience. This chapter explores how induced pressure pain influences cardiac activity elicited via the autonomic nervous system. We aim to uncover signatures in cardiac responses via personalized analysis of the frequencies and the timings of the heart's inter-beat-interval. Autonomic responses such as cardiac activity serve as inevitable processes, which cannot be volitionally controlled—they exhibit a narrow range of dynamics, helping provide robust signatures of the body's responses to pain/distress. We find that pain elicits shifts in the heart rate variability metrics of the cardiac signal, alluding to changes in sympathetic and parasympathetic nervous system activation. Unique relationships are also observed between metrics obtained from the physiological data and self-reported pain ratings. The implications of this work are discussed in the context of precision medicine with possible applications in clinical populations such as autism.

Keywords: cardiac, pain, distress, sympathetic, biometric, ECG, HRV, autism, ASD

1. Introduction

Pain and distress are intrinsically undesirable experiences that are implicated in a variety of physical and mental illnesses. Pain stands at the intersection of multiple health crises, contributing to the opioid epidemic, health disparities, disability, and chronic pain [1, 2]. At least 125 million Americans suffer from acute or chronic pain, and this epidemic has been the root cause of the opioid crisis that arose in the late 1990s [2]. The increased use and misuse of opioids has led to over 47,000 deaths in the United States between 2013 and 2017 alone [3]. Thus, gaining a complete understanding of the neurobiological underpinnings of pain can lead to the most effective solutions to this epidemic [2].

Pain is yet to be explored and digitally characterized in terms of its effects on the autonomic nervous system (ANS). Aside from potential tissue damage, pain is associated with sensory, motor, cognitive, and social components [1]. Investigating pain thus requires multidisciplinary approaches that can integrate insights from psychology (behavior, cognition, sensation, perception), neuroscience (nervous system physiology), and psychiatry (social/clinical research). An objective and noninvasive assessment of pain is also yet to be discovered and utilized in the clinical realm.

Traditional self-report techniques to assess pain are useful and convenient in the clinical realm however, they should be complemented with more objective approaches. Current pain assessments rely on surveys and questionnaires such as numerical rating scales, illustrative visual analog scales, and verbal rating scales which rely on semantic descriptors such as 'moderate' and 'severe' [4, 5]. Such assessments often assume the individual has the capacity to understand and verbalize mental/emotional states, making them disadvantageous for minimally verbal individuals or those with disabilities or neurodevelopmental disorders such as Autism Spectrum Disorder (ASD). Autistic children may experience difficulties in expressing their internal emotional states (hunger, pain, fatigue, etc.), leading to increased stress, tantrums/outbursts, and meltdowns. While external behavioral measures may be helpful in understanding such states, internal states may easily be masked or differently expressed across individuals, leading to interpretation errors. Thus, evaluating nervous system physiology in such populations can greatly enhance approaches that only rely on self-report measures and/or observing external behaviors.

In this work we evaluate the effect of pain on autonomic cardiac regulation, an inevitable process that cannot be consciously controlled during experimental tasks. The current study ultimately aims to develop digital biomarkers that can be used to detect pain and distress from cardiac activity elicited via the autonomic system. With the advent of wearable sensing technologies, it is possible to track physiologically relevant signals (electrocardiography/ECG) to help assess an individual's autonomic states. This study will utilize a multifaceted approach that investigates the effects of pain on cardiac reactivity in relation to self-reports of pain and pain sensitivity. With ECG sensors, we track the dynamics of heart signals and characterize how pain influences autonomic regulation. Pain-related biosignatures obtained via wearable sensors as the person

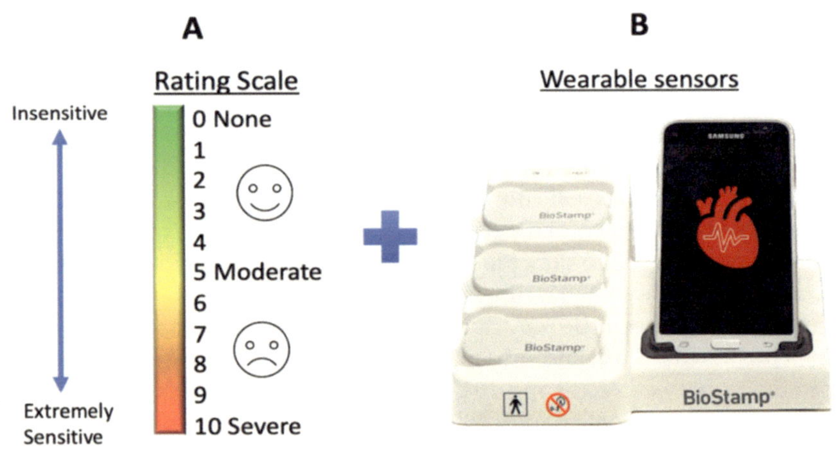

Figure 1.
Integration of subjective and objective metrics to assess states of pain and distress. (A) Self-reports of pain and pain sensitivity levels are assessed by numeric rating scales traditionally used in healthcare settings to assess pain. (B) Wearable sensors that can track heart signals serve as a proxy for autonomic nervous system activity.

experiences physical pain are compared to the results of pain assessments self-reported by the individual. This integrative approach (**Figure 1**) leverages information from the autonomic systems to help in developing a clearer psychophysiological understanding of pain and ultimately aims to create robust techniques to assess pain and distress in those who have difficulty expressing it and in the general population.

2. Cardiac signals as a proxy for autonomic nervous system (ANS) regulation

Heart activity is under the dynamic control of the sympathetic cardiac nerves and the parasympathetic vagus nerve via the autonomic branches of the peripheral nervous system (NS). The ANS is largely responsible for maintaining the body's overall homeostasis [5]. The sympathetic NS works to increase heart rate while the parasympathetic NS serves as the brakes that turn the cardiac activity back to normal functioning. Sensory neurons between the brain, spinal cord, and cardiac muscles engage in continuous feedback loops, consistently influencing each other via re-afferent signals (**Figure 2A**).

Fluctuations in sympathetic and parasympathetic activity can allude to unique physiological responses related to stress and anxiety. Exposure to painful stimuli and/or discomfort and distress can activate the sympathetic NS which elicits the excitatory fight-or-flight response [6]. Previous studies on stress and autonomic responses such as cardiac reactivities provide insight into the physiology of pain sensation [7, 8]. Heart rate variability (HRV) analyses have proven reliable and advantageous in evaluating autonomic functions in this regard [9, 10]. HRV metrics represent the various statistics of the inter-beat- interval (IBI), the timing between beats in a cardiac signal (**Figure 2B**). HRV is widely used to evaluate sympathetic and parasympathetic

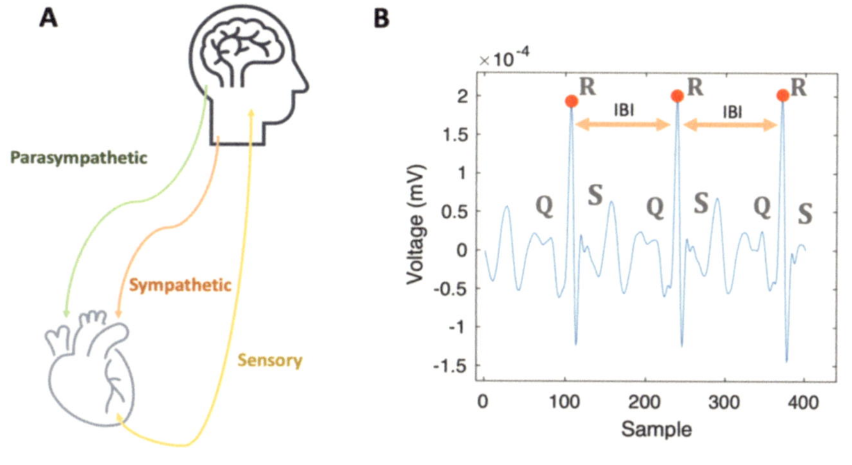

Figure 2.
Cardiac responses as a proxy for autonomic nervous system function. (A) The sympathetic nervous system is responsible for increasing cardiac contractions during states of distress/anxiety while the parasympathetic nervous system serves as the brakes on the sympathetic system, modulating cardiac activity during resting states. Parasympathetic activity is guided by the vagal nerve, which governs heart rate variability (HRV) and allows for adaptive behaviors. Sensory neurons via the spinal cord allow for two-way communication between the heart and the brain. (B) Cardiac signals exhibit a unique QRS pattern where the timing between R-peaks (IBI) can serve as a proxy for sympathetic and parasympathetic activity.

NS activity via various time and frequency-domain parameters. Improper balance between these two systems is often associated with cardiac pathologies such as strokes and heart attacks [9].

Previous work on HRV and stress have shown that when humans experience mental/physical strain, the parasympathetic NS's control over the heart decreases while sympathetic NS activity increases [11–13]. In such studies, physical strain was induced by having subjects perform intensive exercises [12], and mental strain was induced by asking subjects to solve difficult puzzles or arithmetic problems [13]. From these studies we find a clear interaction between the ANS and the nociceptive system as pain may often induce both mental and/or physical strain and distress [5, 6].

2.1 Autonomic dysregulation: social and emotional components

The autonomic nervous system plays an important role in socio-emotional learning and control [14]. Balanced vagal tone via the parasympathetic NS (responsible for modulating heart-rate) allows for swift engagement and disengagement with people, which is important for building social communication skills [15]. Social skills development during early childhood predicts a range of positive outcomes (in communication, assertiveness, role transitions, etc.) along with the formation and management of family and peer relationships throughout the lifespan [16]. Increased vagal tone is also associated with higher facial expressivity levels [17]. Decreased vagal tone can lead to dysregulated heart-rate (HR) modulation which may in turn lead to social interaction difficulties [15]. Imbalanced autonomic activity can also contribute to socio-emotional dysregulation during dyadic interactions in children diagnosed or at-risk for psychopathologies [18].

Symptoms of ASD are proposed to be associated with autonomic dysfunctions. Previous studies show that children with ASD and Intellectual Disability (ID) exhibit low parasympathetic activity during high anxiety conditions [19]. Autonomic dysregulation is also apparent in autistic children compared to typically developing (TD) controls [20]. When comparing the cardiac and electrodermal activity of autistics and neurotypicals, those with ASD exhibited dampened HR reactivity and skin conductance responses to visual and auditory social stimuli (face images and speech sounds) and during social interactions (role play) [21, 22]. More recent work demonstrated that the non-linear metrics of HRV show decreased autonomic modulation in autistic individuals compared to controls during resting conditions [23]. During facial expression tasks where subjects were asked to draw, interpret, and recognize different emotions, the ASD group showed lower parasympathetic modulation compared to controls, alluding to the elicitation of cognitive stress. Such autonomic dysregulation was also correlated with autism severity [24]. During social attention tasks, autistic subjects similarly showed reduced parasympathetic modulation, suggesting that autonomic dysregulation may underly social deficits in ASD [25].

Such findings are in line with the Polyvagal Theory which suggests that social behaviors may arise from the autonomic nervous system, with efficient vagal/parasympathetic control preventing sympathetic overactivation and thus contributing to better socio-emotional skills [26]. In ASD, the 'vagal brakes' on the sympathetic system may be compromised, leading to sympathetic hyperarousal and increased distress, which may impair behavioral adaptation/control and the ability to satisfactorily reciprocate social interactions [27]. Thus, assessing autonomic modulation may prove useful in understanding social–emotional responses, adaptive behaviors, and ultimately in screening and tracking symptoms associated with ASD.

DOI: http://dx.doi.org/10.5772/intechopen.111890

3. Experimental approach: autonomic biomarkers

In this work, we explore the cardiac activity of neurotypical (TD) and autistic (ASD) individuals. Autonomic responses were proxied via the electrical activity of the heart. Electrocardiographic (ECG) activity was captured via wearable biosensors placed on the chest at the standardized lead II position via gel adhesives.

At the beginning of the study, TD participants were asked to rate their perceived pain sensitivity (PPS) relative to other people on a scale of 0–10, where zero represents complete insensitivity and a 10 represents extreme sensitivity [28].

Participants were seated at a table where they performed a Resting Task under control and experimental conditions. In the control condition of the study, the participants performed the Resting Task by sitting in a relaxed position and avoiding excess movement for about two minutes. This task was used to establish baseline autonomic NS activity as no movement or cognitive effort was required. During the experimental/pain condition of the study (only conducted with TD subjects), sustained pressure pain was introduced via a manual blood pressure cuff. This pain induction method serves as a modified version of the submaximal effort tourniquet test [28] which is found to mimic pathological pain [29]. In this procedure, the blood pressure cuff (standard sphygmomanometer/tourniquet) was placed around the non-dominant arm of the participant (above the elbow) and was gradually inflated to a pressure level of about 200 mmHg [30]. The cuff was inflated at this level for the entirety of the task and was deflated right after task completion. Right before cuff deflation, TD subjects verbally reported their pain level using a Numeric Rating Scale (NRS) ranging from 1 to 10, where a 1 indicates minimal to no pain, 4–6 indicates moderate pain, and a 10 indicates extreme pain [31]. Participants in the ASD group performed the Resting Task only under the control condition.

3.1 Analytical approach to assess cardiac activity

Electrocardiographic (ECG) data typically includes consecutive QRS complexes representing each heartbeat in the cardiac signal (**Figure 2B**). The R-peaks (sharp spikes) within QRS complexes are traditionally used to assess the timing between consecutive heartbeats (known as the R-R or inter-beat interval). Accurately detecting R-peaks is essential for assessing the fluctuations in the inter-beat interval (IBI) signal and in computing various heart rate variability (HRV) parameters. ECG signals may easily be corrupted by various artifacts such as baseline wandering and electrode movement [32]. To clean the raw ECG data, signal-processing filters were used to minimize excess noise/frequencies outside the range of a typical ECG recording [33]. After preprocessing of the ECG data, R-peaks were detected via a simple peak detection algorithm in MATLAB (MathWorks) software, and the IBI signal was obtained by computing the time between consecutive R-peaks. The statistics of the IBI signal were then evaluated via various frequency and time-domain metrics.

3.2 Cardiac activity metrics of the autonomic system

Changes in heart rate are the result of autonomic control via the sympathetic (excitatory) and parasympathetic (inhibitory) nervous systems (NS), which are informative in assessing how pain and distress influence the autonomic system [5]. HRV describes fluctuations in instantaneous heart rate (the oscillations between two consecutive heartbeats), where greater variability often reflects enhanced vagal tone (heart rate regulation). The activities of the parasympathetic and sympathetic NS can be inferred from the Power Spectral Density (PSD) and Poincaré plots of the HRV signal (**Figure 3**). The high frequency (HF) component of the PSD (150–400 mHz range) is associated with parasympathetic activity and a general decrease in heart rate [34]. The low frequency (LF) component of the PSD (40–150 mHz range) is associated with sympathetic NS activity and blood pressure control [34]. The LF/HF ratio reflects the sympatho-vagal balance – the contribution of the sympathetic NS in controlling the heart compared to the parasympathetic NS [9]. Increases in the ratio between the LF and HF components (LF/HF ratio) have been previously associated with stress and intense exercise [12, 13]. The LF and HF components were computed by integrating the PSD over the associated frequency range.

Poincaré plots are also used to assess sympathetic and parasympathetic activation via time-domain metrics. Pioncaré plots serve as a geometrical and nonlinear method to assess the dynamics of HRV and are formed via a scatter of the IBI interval against the preceding IBI interval [35]. The width of the scatter is used to determine the SD1 parameter, which reflects parasympathetic NS activity and is correlated with HF power [36, 37]. The length of the scatter is used to determine the SD2 parameter, which reflects sympathetic NS activity and is correlated with LF power [36].

The PhysioNet Cardiovascular Signal Toolbox was also used to assess the ECG time series. This open-source toolbox is designed to assess HRV via standardized algorithms [38]. With this toolbox we windowed the IBI series and obtained a distribution of LF, HF, LF/HF, SD1, SD2, and SD2/SD1 parameters. To better visualize which frequencies (in the LF and HF ranges) dominated the cardiac signal across time, continuous wavelet transforms (CWT) or magnitude scalograms were used to visualize and evaluate temporal changes in frequency power and provide a personalized assessment of the autonomic activity (**Figure 4A**).

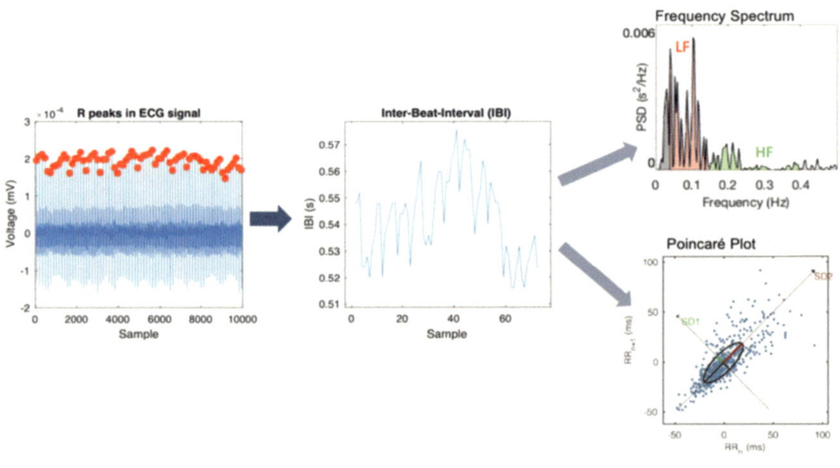

Figure 3.
Analytical pipeline to assess autonomic activity via cardiac signals. The timing between R peaks (red markers) of the original ECG data are extracted to obtain the IBI signal. The IBI signal is then assessed in the frequency domain (power spectrum) to obtain LF (sympathetic) and HF (parasympathetic) components. The same IBI signal is assessed in the time domain (Poincaré plot) to obtain the SD2 (sympathetic) and SD1 (parasympathetic) parameters.

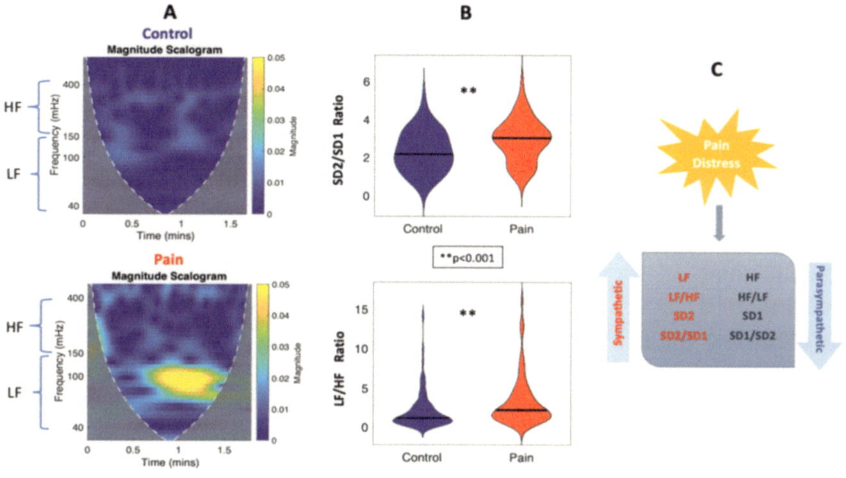

Figure 4.
Cardiac responses across control and pain conditions in TD subjects. (A) CWT plots (or magnitude scalograms of the frequencies present in the heart signal across time) demonstrate higher magnitudes in the LF range corresponding to sympathetic nervous system activation in the pain compared to the control condition. (B) Violin plots from data accumulated across subjects demonstrate that the LF/HF and SD2/SD1 ratios (indicative of sympathetic NS arousal and/or parasympathetic NS inhibition) of the control vs. pain condition arise from different distributions, with the median for these ratios being higher for the pain condition. (C) Schematic of the effects of pain and pain-related distress on the autonomic system and the corresponding changes observed in the HRV metrics in the time (SD2 and SD1) and frequency (LF and HF) domains.

3.3 Integrating cardiac biometrics with self-report data

To explore the relationship between self-reported responses and the cardiac biometric data, scatter plots were made comparing each subject's ratings against the absolute difference between the HRV metrics (obtained from the experimental and control conditions). Numeric scale pain ratings of the experimentally induced pressure pain were also compared to the perceived pain sensitivity (PPS) ratings. Scatterplots were used to assess possible relationships between PPS ratings and changes in the biometrics obtained from the cardiac signal. Such methods allow us to evaluate the correspondence between physiological metrics and common psychological assessments of pain.

4. Results: autonomic responses of distress

4.1 Heart rate variability (frequency + time-domain metrics)

HRV results were compared across control and pain conditions for TD subjects (**Figure 4**). Frequency-domain analysis of the IBI data demonstrated that the pain condition often elicited an increase in LF power (corresponding to sympathetic NS activation) and/or a decrease in HF power (corresponding to parasympathetic NS activity) for TD subjects. Such frequency changes can be visualized qualitatively across the entirety of the task via CWT plots (**Figure 4A**). The LF/HF and SD2/SD1 ratios (where an increase represents sympathetic NS activation or parasympathetic NS inhibition) were also computed across the control and pain conditions. Violin plots demonstrated changes in the shape of the probability density of these parameters across participants (**Figure 4B**). For the pain condition, there was a general increase in both ratios based on data accumulated across TD subjects. Nonparametric Kruskal-Wallis tests indicated that the LF/HF and SD2/SD1 ratios across the control and pain conditions come from significantly different distributions: $\chi^2(1,175) = 14.37$, $p < 0.001$ and $\chi^2(1,179) = 12.44$, $p < 0.001$, respectively.

4.1.1 The unique case of a subject with chronic pain

Heart data analyses for a subject known to experience chronic pain led to unique findings compared to the remaining participants. The CWT plots of this subject consistently exhibited high LF power specifically in the 100–150 mHz range (**Figure 5**). This pattern was observed across both control and pain conditions. It is important to note that while this participant did experience the experimentally induced pressure sensation during the pain condition, they were accustomed to experiencing a consistent level of pain throughout their daily life. These results emphasize the importance of analyzing at the biophysical data in a personalized manner before calculating summary statistics or assessing trends based on the entire sample.

4.2 Self-reported pain ratings and HRV parameters

When exploring the self-reported measures, we find that perceived pain sensitivity (PPS) generally corresponded with the pain ratings reported during the study (**Figure 6A**). This indicated that participants could accurately approximate their pain sensitivity levels compared to others. To assess the relationship between objective and self-reported measures, we compared the HRV parameters of the cardiac signal with each subject's ratings during the pain condition of the Resting Task. The absolute difference (Diff) in the SD2 parameter (indicative of sympathetic NS activity) was computed to assess how much each subjects' cardiac signatures during the pain experience deviated from those of the control condition (SD2 Diff). This deviation in the SD2 parameter appeared to positively correlate with self-reported pain ratings and

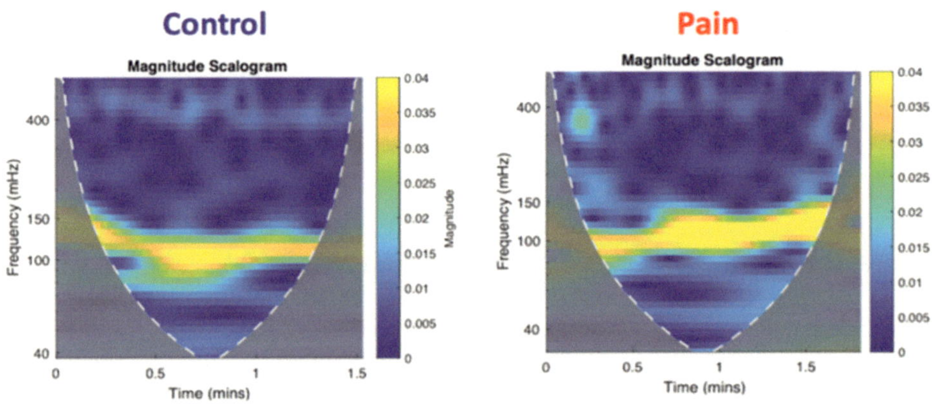

Figure 5.
Cardiac activity of a participant with chronic pain. CWT plots consistently showed high magnitude in the 100–150 mHZ range (corresponding to LF power) regardless of whether pressure pain was or was not (control condition) introduced during the resting task.

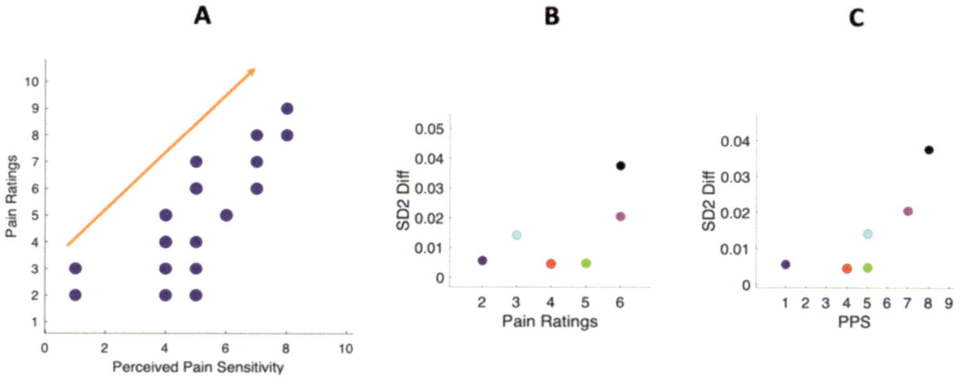

Figure 6.
Integrating self-reported measures with autonomic HRV biometrics. (A) Participants with higher perceived pain sensitivity (PPS) levels typically reported the induced pressure pain to be of a higher intensity on the 1–10 numeric rating scale. (B) The absolute difference in the SD2 biometric between the control and pain condition (SD2 Diff) generally corresponded with higher self-reported pain ratings and perceived pain sensitivity levels.

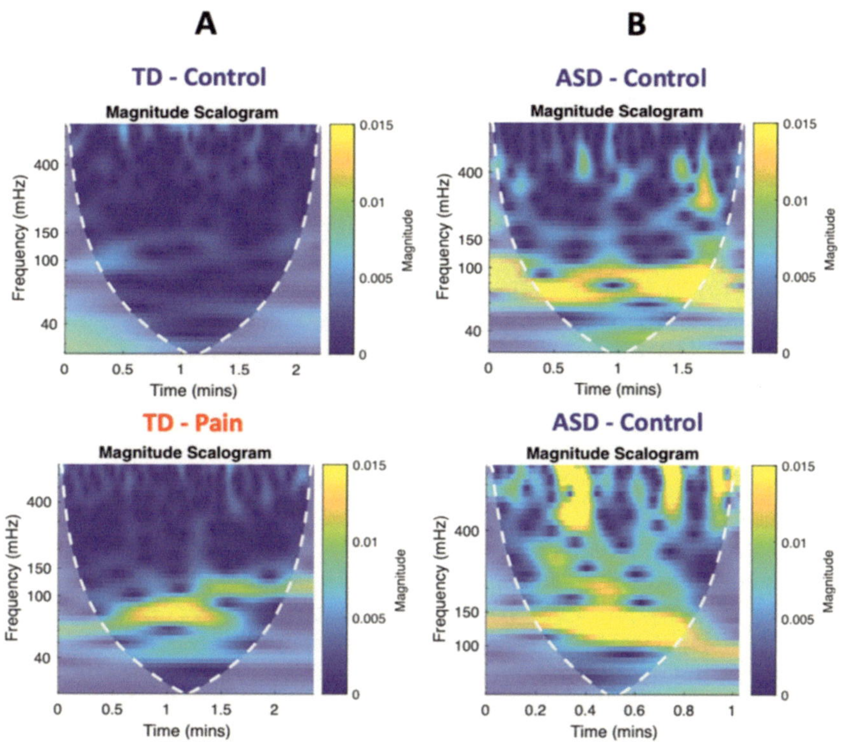

Figure 7.
Comparing cardiac responses of TD and ASD participants. (A) Cardiac activity of neurotypical (TD) participants during the pain condition exhibited higher power in the LF range of the magnitude scalogram compared to the control condition where no pressure pain was induced. (B) For ASD participants at baseline, we observe a similar pattern of higher power in the LF range that mimics what is observed in TD participants during the pain condition.

perceived pain sensitivity (**Figure 6B-C**). This indicates that it is possible to elucidate relationships between objective and subjective measures of pain sensation.

4.3 Autonomic responses in ASD subjects at baseline

When assessing the cardiac activity of autistic participants during the control condition of the Resting Task, we see a pattern in the CWT magnitude scalograms that mimics what was observed in TD participants at baseline (**Figure 7**). This indicated that autistic individuals may be experiencing sympathetic hyperarousal or have dampened parasympathetic nervous system regulation at baseline.

5. Discussion

The goal of this work was to assess changes in autonomic nervous system responses when the body experienced physical pain. We aimed to determine whether pain could be objectively characterized via heart rate variability metrics and how these metrics would compare to self-reported measures of pain. The pressure pain's influence on cardiac activity was apparent in the HRV metrics of the ECG signal. In the TD group, the induced pain led to increases in LF power. For most subjects, we also saw a corresponding decrease in HF power, which ultimately led to an increased LF/HF ratio. Such changes in the frequency-domain metrics suggest that the pressure pain led to sympathetic nervous system activation [25]. Interestingly, when assessing the cardiac activity of a participant who experienced chronic pain, we detected a consistent band of LF power (sympathetic activity) dominating the signal across both the control and pain conditions. This may indicate that the subject's chronic pain elicits a cardiac response that is impervious to the experimentally induced pressure pain. The finding that this subject's baseline cardiac activity resembles that of TD subjects under the pain condition helps provide some external validity to the pain induction procedure and provides further evidence for how pain can lead to increases in LF power and the LF/HF ratio. When evaluating HRV metrics in the time-domain via Poincaré plots, we generally observed an increase in the SD2, a decrease in the SD1, and an increase in the SD2/SD1 ratio, each of which are associated with sympathetic NS overdrive [35]. We find similar patterns of sympathetic hyperarousal in autistic individuals at baseline.

In this work, the physiological HRV metrics also complemented self-report measures. In general, we found that subjects have an accurate perception of their pain sensitivity level, as their PPS ratings appeared to positively correlate with their self-reported pain ratings. When assessing the relation between HRV metrics of the cardiac signal and the self-reported responses, we observed that higher deviations in the SD2 between the baseline and pain condition corresponded to higher PPS and numeric pain ratings. Such findings indicate that objective and nonlinear measures of HRV such as the SD2 parameter can be informative in understanding an individual's pain levels and their general pain sensitivity.

5.1 Study implications and future work

With this work we can begin to understand the relationship between pain, psychological responses, and physiological activity. This study demonstrated that the influence of pain on the body can be characterized via the statistics obtained from cardiac signals and that such biometrics can inform current subjective approaches. The findings of this study have several clinical implications. Characterizing the effects of sustained pressure on cardiac functioning of healthy individuals can help in the development of accurate and objective digital biomarkers of pain sensation. Such objective assessments are vital to understanding whether and how individuals who may have difficulty communicating their pain – such as those with Autism Spectrum Disorder (1 of 59 in the US), Intellectual Disability, or impaired communication skills – experience pain under normal conditions [39–41]. Individuals with Intellectual Disability experience chronic pain that often goes unnoticed and untreated [41]. Autistic individuals often exhibit higher sensitivity to painful stimuli and generally have trouble communicating their emotional or mental states to others [39, 40]. During states of sympathetic overdrive (as commonly observed in ASD), it is difficult for the parasympathetic system to acquire the predictability needed for expressing internal mental and emotional states and reciprocating social interactions. Thus, there is an urgent need to develop objective methods to characterize and understand pain and distress in such individuals. Detecting any form of autonomic dysregulation may indeed help us understand the hypersensitivities and socio-emotional difficulties observed in ASD. Such assessments can ultimately contribute to early diagnoses and mitigated distress for families and non-speaking populations at large.

Pain is a multi-faceted construct, associated with multiple biological, sensory, cognitive, and social components [1]. Thus, it must be explored via a multidimensional psychophysiological approach. Future work aims to explore the sensory-motor and socio-emotional aspects of the pain experience. This can be done by assessing the facial expression and movement activity of participants as they perform motor-cognitive tasks while experiencing pain. While the autonomic system provides a bounty of information about the underlying physiology of pain and distress, we cannot forget about the contribution of the overarching peripheral nervous system that works to guide sensation, perception, decision-making, movement, and overall behavior. Recent work by Ryu and Torres has indeed connected voluntary control of motor output to the autonomic system in neurotypical individuals [42]. This work demonstrated that the heart plays a vital role in agency, highlighting the delicate balance between autonomy and control. From this previous work we learned that the cardiac signal leads the motor signal when a movement is intended but lags it when the movement is unintended [42]. Besides differentiating between deliberate and spontaneous motions, the cardiac code can help us begin the path of characterizing and distinguishing different types of afferent feedback, including those from kinesthetic and somatic pain signals. Further evaluating the autonomic system can help us deconvolve the continuous efferent stream of voluntary movements from the afferent consequences that they themselves cause. Understanding such relations will help us derive causal mechanisms of the nervous systems, beyond mere correlations. Such work highlights the importance of exploring the peripheral nervous system (including the autonomic branches) as a whole, as it can play key roles in the multi-faceted nature of pain and distress.

6. Conclusions

This work provides an innovative approach to better understand the mechanisms by which experimentally induced pain – that mimics pathological pain [28] – influences the autonomic nervous system via evaluating cardiac signals. From this work, we learn that pain can interfere with autonomic regulation, eliciting sympathetic overdrive. The cardiac reactivities also appear to correspond with self-reports of pain and pain sensitivity. The observed patterns of autonomic dysregulation (sympathetic hyperarousal) during the physical pain experience in TD individuals mimic the cardiac responses observed in ASD participants at baseline. The unique results observed in ASD and chronic pain subjects highlight the importance of a personalized approach to assessing data. Such methodologies lend themselves to the Precision Medicine platform which helps inform the development of personalized treatments [43]. Our psychophysiological approach can ultimately help create robust techniques to detect pain and aid in the development of personalized interventions that are tailored to each individual's autonomic functioning. The added convenience of using wearable sensors makes this technique flexible and translatable for use in healthcare settings. The digital biometrics explored in this work open a new realm of research that can help us scientifically understand and characterize pain in a variety of neurodevelopmental disorders and in those with communication disabilities. Ultimately, such research can lead to new methods of identifying and alleviating pain and distress, improving the quality of life of individuals across the globe.

Additional information

Portions of this book chapter are derived from the thesis project titled "Characterization of Psychophysiological Responses to Pressure Pain" authored by Mona Elsayed, which is available on the Rutgers University repository platforms, dated October 2021. The thesis work has not been peer-reviewed nor published elsewhere.

Acknowledgements

We thank all members of the Sensory Motor Integration Lab (SMIL), and all the families and participants who made this work possible. This work was supported by the Nancy Lurie Marks Family Foundation Development Career Award to EBT and the New Jersey Governor's Council for Research and Treatment of Autism to EBT (CAUT14APL018).

Conflict of interest

The authors declare no conflict of interest.

DOI: http://dx.doi.org/10.5772/intechopen.111890

Author details

Mona Elsayed* and Elizabeth Barbara Torres
Rutgers University, Piscataway, NJ, USA

*Address all correspondence to: mona.elsayed@rutgers.edu

References

[1] Nugraha B, Gutenbrunner C, Barke A, Karst M, Schiller J, Schäfer P, et al. The IASP classification of chronic pain for ICD-11: Functioning properties of chronic pain. Pain. 2019;**160**(1):88-94

[2] Skolnick P. The opioid epidemic: Crisis and solutions. Annual Review of Pharmacology and Toxicology. 2018;**58**(1):143-159

[3] Seth P, Scholl L, Rudd RA, Bacon S. Overdose deaths involving opioids, cocaine, and psychostimulants—United States, 2015-2016. American Journal of Transplantation. 2018;**18**(6):1556-1568

[4] Kvistgaard Olsen J, Fener DK, Wæhrens EE, Wulf Christensen A, Jespersen A, Danneskiold-Samsøe B, et al. Reliability of pain measurements using computerized cuff algometry: A DoloCuff reliability and agreement study. Pain Practice. 2017;**17**(6):708-717

[5] Cardinali DP. Autonomic Nervous System: Basic and Clinical Aspects. Switzerland: Springer International Publishing AG; 2017

[6] Schlereth T, Birklein F. The sympathetic nervous system and pain. Neuromolecular Medicine. 2008;**10**:141-147

[7] Koenig J, Jarczok MN, Ellis RJ, Hillecke TK, Thayer JF. Heart rate variability and experimentally induced pain in healthy adults: A systematic review. European Journal of Pain. 2014;**18**(3):301-314

[8] Shaffer F, Ginsberg JP. An overview of heart rate variability metrics and norms. Frontiers in Public Health. 2017;**5**:258

[9] Wang Y, Hensley MK, Tasman A, Sears L, Casanova MF, Sokhadze EM. Heart rate variability and skin conductance during repetitive TMS course in children with autism. Applied Psychophysiology and Biofeedback. 2016;**41**:47-60

[10] Kim H, Cheon E, Bai D, Lee YH, Koo B. Stress and heart rate variability: A meta-analysis and review of the literature. Psychiatry Investigation. 2018;**15**(3):235-245. DOI: 10.30773/pi. 2017.08. 17

[11] Hunt KJ, Fankhauser SE. Heart rate control during treadmill exercise using input-sensitivity shaping for disturbance rejection of very-low-frequency heart rate variability. Biomedical Signal Processing and Control. 2016;**30**:31-42

[12] Rani P, Sims J, Brackin R, Sarkar N. Online stress detection using psychophysiological signals for implicit human-robot cooperation. Robotica. 2002;**20**(6):673-685

[13] Mulkey SB, du Plessis AJ. Autonomic nervous system development and its impact on neuropsychiatric outcome. Pediatric Research. 2019;**85**(2):120-126

[14] Porges SW. The Polyvagal Theory: Neurophysiological Foundations of Emotions, Attachment, Communication, and Self-Regulation (Norton Series on Interpersonal Neurobiology). New York, NY: WW Norton & Company; 2011

[15] Ainsworth MD, Bell SM. Some contemporary patterns of mother-infant interaction in the feeding situation. Stimulation in Early Infancy. London and New York: Academic Press; 1969

[16] Cole PM, Zahn-Waxler C, Fox NA, Usher BA, Welsh JD. Individual differences in emotion regulation and behavior problems in preschool children.

Journal of Abnormal Psychology. 1996;**105**(4):518

[17] Shahrestani S, Stewart EM, Quintana DS, Hickie IB, Guastella AJ. Heart rate variability during social interactions in children with and without psychopathology: A meta-analysis. Journal of Child Psychology and Psychiatry. 2014;**55**(9):981-989

[18] Moskowitz LJ, Mulder E, Walsh CE, McLaughlin DM, Zarcone JR, Proudfit GH, et al. A multimethod assessment of anxiety and problem behavior in children with autism spectrum disorders and intellectual disability. American Journal on Intellectual and Developmental Disabilities. 2013;**118**(6):419-434

[19] Anandhi B, Jerritta S, Anusuya IG, Das H. Time domain analysis of heart rate variability signals in valence recognition for children with autism Spectrum disorder (ASD). IRBM. 2022;**43**(5):380-390

[20] Hirstein W, Iversen P, Ramachandran V. Autonomic responses of autistic children to people and objects. Proceedings of the Royal Society of London. Series B: Biological Sciences. 2001;**268**(1479):1883-1888

[21] Smeekens I, Didden R, Verhoeven EW. Exploring the relationship of autonomic and endocrine activity with social functioning in adults with autism spectrum disorders. Journal of Autism and Developmental Disorders. 2015;**45**:495-505

[22] Julu PO, Kerr AM, Apartopoulos F, Al-Rawas S, Engerström IW, Engerström L, et al. Characterisation of breathing and associated central autonomic dysfunction in the Rett disorder. Archives of Disease in Childhood. 2001;**85**(1):29-37

[23] Edmiston EK, Jones RM, Corbett BA. Physiological response to social evaluative threat in adolescents with autism spectrum disorder. Journal of Autism and Developmental Disorders. 2016;**46**:2992-3005

[24] Gonzaga CN, Valente HB, Ricci-Vitor AL, Laurino MJ, Santos LA, Stoco-Oliveira MC, et al. Autonomic responses to facial expression tasks in children with autism spectrum disorders: Cross-section study. Research in Developmental Disabilities. 2021;**116**:104034

[25] Porges SW. Orienting in a defensive world: Mammalian modifications of our evolutionary heritage. A polyvagal theory. Psychophysiology. 1995;**32**(4):301-318

[26] Porges SW. The polyvagal theory: Phylogenetic contributions to social behavior. Physiology & Behavior. 2003;**79**(3):503-513

[27] Ruscheweyh R, Marziniak M, Stumpenhorst F, Reinholz J, Knecht S. Pain sensitivity can be assessed by self-rating: Development and validation of the pain sensitivity questionnaire. Pain. 2009;**146**(1-2):65-74

[28] Pertovaara A, Nurmikko T, Pöntinen PJ. Two separate components of pain produced by the submaximal effort tourniquet test. Pain. 1984;**20**(1): 53-58

[29] Sternbach RA, Deems LM, Timmermans G, Huey LY. On the sensitivity of the tourniquet pain test. Pain. 1977;**3**(2):105-110

[30] Moore PA, Duncan GH, Scott DS, Gregg JM, Ghia JN. The submaximal effort tourniquet test: Its use in evaluating experimental and chronic pain. Pain. 1979;**6**(3):375-382

[31] Krebs EE, Carey TS, Weinberger M. Accuracy of the pain numeric rating scale as a screening test in primary care. Journal of General Internal Medicine. 2007;**22**:1453-1458

[32] Kameenoff J, Kameenoff J. Signal processing techniques for removing noise from ECG signals. Biomedical Engineering and Research. 2017;**1**(1):1

[33] Fedotov AA. Selection of parameters of bandpass filtering of the ECG signal for heart rhythm monitoring systems. Biomedical Engineering. 2016;**50**:114-118

[34] Laborde S, Mosley E, Thayer JF. Heart rate variability and cardiac vagal tone in psychophysiological research–recommendations for experiment planning, data analysis, and data reporting. Frontiers in Psychology. 2017;**8**:213

[35] Brennan M, Palaniswami M, Kamen P. Do existing measures of Poincare plot geometry reflect nonlinear features of heart rate variability? IEEE Transactions on Biomedical Engineering. 2001;**48**(11):1342-1347

[36] Hsu CH, Tsai MY, Huang GS, Lin TC, Chen KP, Ho ST, et al. Poincaré plot indexes of heart rate variability detect dynamic autonomic modulation during general anesthesia induction. Acta Anaesthesiologica Taiwanica. 2012;**50**(1):12-18

[37] Goit RK, Ansari AH. Reduced parasympathetic tone in newly diagnosed essential hypertension. Indian Heart Journal. 2016;**68**(2):153-157

[38] Vest AN, Da Poian G, Li Q, Liu C, Nemati S, Shah AJ, et al. An open source benchmarked toolbox for cardiovascular waveform and interval analysis. Physiological Measurement. 2018;**39**(10):105004

[39] Allely CS. Pain sensitivity and observer perception of pain in individuals with autistic spectrum disorder. The Scientific World Journal. 2013;**2013**:916178-20

[40] Ely E, Chen-Lim ML, Carpenter KM, Wallhauser E, Friedlaender E. Pain assessment of children with autism spectrum disorders. Journal of Developmental & Behavioral Pediatrics. 2016;**37**(1):53-61

[41] McGuire BE, Daly P, Smyth F. Chronic pain in people with an intellectual disability: Under-recognised and under-treated? Journal of Intellectual Disability Research. 2010;**54**(3):240-245

[42] Ryu J, Torres E. The autonomic nervous system differentiates between levels of motor intent and end effector. Journal of Personalized Medicine. 2020;**10**(3):76

[43] Hawgood S, Hook-Barnard IG, O'Brien TC, Yamamoto KR. Precision medicine: Beyond the inflection point. Science Translational Medicine. 2015;**7**(300):300ps17

Section 3

Neural Control in Unique Systems

Chapter 5

Sympathetic Innervation of the Mammalian Pineal Gland: Its Involvement in Ontogeny and Physiology, and in Pineal Dysfunction

Martin Avila, Carlos L. Freites, Elena Vásquez, Juan B. Amiotti, Janina Borgonovo and Estela M. Muñoz

Abstract

In mammals, the melatonin-producing pineal gland (PG) receives sympathetic innervation from the superior cervical ganglia (SCG). This chapter describes the role of this innervation on the PG's ontogeny and rhythmic function, along with consequences to physiology when this regulation is disrupted. The PG and the SCG are components of the circadian timing system (CTS). Therefore, the overall CTS is described, including its oscillatory basis, its synchronization to the light: dark (L:D) cycles, and the dissemination of timing cues to all cells throughout the body. Pineal cellular composition and heterogeneity, cell-cell interactions, and the molecular mechanisms involved in the circadian rhythm of melatonin (MEL), are discussed. The SCG's bilateral placement among surrounding anatomical landmarks, as well as their afferent and efferent connections, are described and illustrated. In addition, the SCG-related surgical models and the state-of-the art technology used to investigate the connection between SCG and PG are presented. Perspectives and gaps in our understanding are also discussed. We hope this chapter inspires readers to delve deeper into the field of the pineal gland and its main messenger, melatonin, as well as MEL's impact in health and disease, including as a remedial therapy.

Keywords: hormone, melatonin, pineal gland, pinealocyte, sympathetic innervation, superior cervical ganglia, norepinephrine, ontogeny, physiology, dysfunction

1. Introduction

In mammals, melatonin (MEL) is a circadian hormone that is released at high levels into the bloodstream and into the cerebrospinal fluid (CSF) at night, but then drops off to negligibly low levels throughout the daytime [1–3]. Almost all the body's cells respond to this timing signal. MEL and other circadian cues orchestrate physiology in a rhythmic manner that impacts organ function, tissue healing and rejuvenation, and growth, as well as cognition, motivation, behavior, adaptation, and survival [4]. Taking melatonin supplements is growing in popularity, mainly to augment naturally produced MEL levels and as a sleep aide, but also for its powerful antioxidant, anti-inflammatory, free-radical scavenging, and neuroprotective properties [5]. In mammals, circulating MEL is primarily synthesized by the pineal gland (PG), under direct control of sympathetic innervation stemming from the superior cervical ganglia (SCG) [6, 7]. The PG and the SCG are part of an endogenous circadian timing system (CTS) that synchronizes the whole organism to the environmental light: dark cycles (L:D; *Zeitgeber*). In this chapter, we present foundational and current knowledge about how the pineal gland is controlled by the sympathetic nervous system (SNS), with regard to its ontogeny and normal physiology, as well as under dysfunctional conditions. We hope this work inspires readers to seek a deeper understanding of the pineal gland and the functional role of its main messenger, melatonin, in both health and disease, as well as remedial therapy.

2. The pineal gland's role in the mammalian circadian timing system

The pineal gland (PG) is a highly vascularized neuroendocrine organ that rhythmically produces melatonin (MEL) [2]. The PG is located in the mid-line of the brain, attached to the roof of the third ventricle (III V) by a short stalk [8, 9]. The PG is positioned deep within the brain of humans, and more superficially in rodents. The mammalian PG is driven by a hierarchical series of oscillators from the photoneuroendocrine system (PNS) (**Figure 1**) [7, 12]. Furthermore, the nocturnal release of MEL by the PG provides downstream circadian synchronization to most cells throughout the body. All these elements taken together comprise the circadian timing system (CTS) [12]. The PNS transduces 24-hour light: dark (L:D) cycle information from the external environment into the circadian pattern of MEL synthesis and secretion. To do this, the multisynaptic PNS senses light using intrinsically photosensitive retinal ganglion cells (ipRGC) in the eye, in coordination with the retinal rods and cones. The ipRGC axons project into the GABAergic neurons in the suprachiasmatic nuclei (SCN) of the hypothalamus. SCN are considered to be the master circadian pacemaker, which synchronizes a complex and widely distributed network of peripheral clocks. Each of these oscillators has its own cell-autonomous circadian clock that is driven by interlocked transcriptional/translational feedback loops (TTFL) of core-clock genes (CG), that in turn regulate the expression of clock-controlled genes (CCG), and ultimately coordinate the timing of many biological processes throughout the body [12–15]. The CG family includes genes that encode either positive or negative regulators, such as CLOCK, BMAL, PERs (Period), and CRYs (Cryptochrome) proteins. During the light phase of the L:D cycle, glutamatergic ipRGC axons activate the SCN neurons. This inhibits the rest of the PNS, including the hypothalamic paraventricular nuclei (PVN), neurons of the intermediolateral columns (IMC) of the spinal cord (SC), and the superior cervical ganglia (SCG), and thus, prevents MEL synthesis and secretion by the PG. During the night phase, the SCN release their inhibition over the circuit, and SCG nerve ends release norepinephrine (NE) into the PG parenchyma [6]. This neurotransmitter binds to specific adrenergic receptors on the pinealocyte (Pc) plasma membrane and regulates key steps in the multienzymatic pathway that results in MEL synthesis.

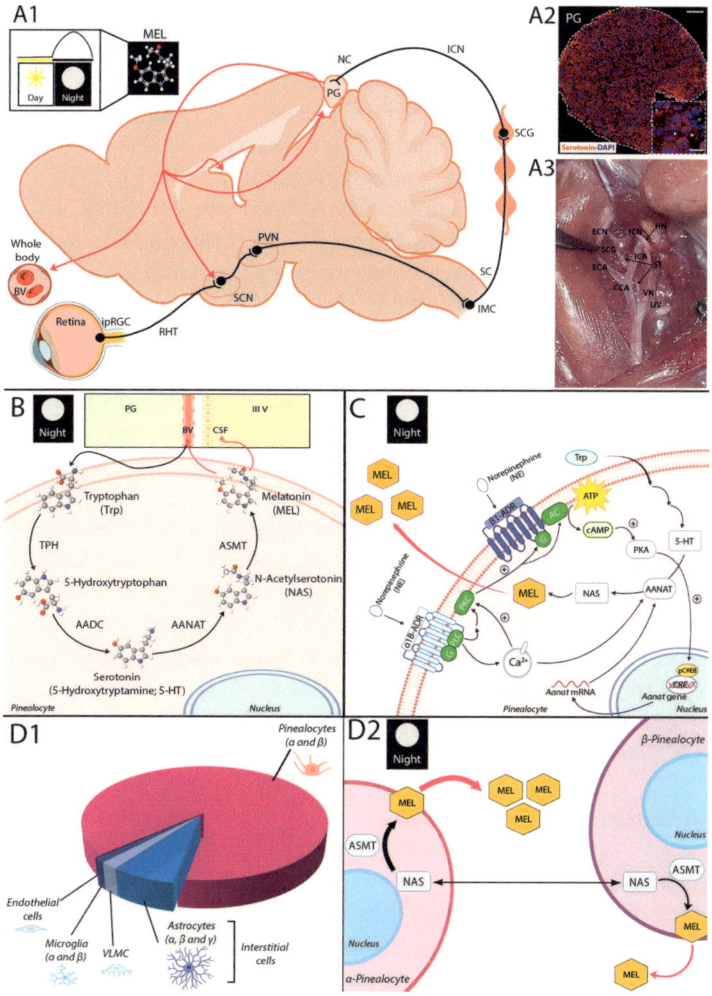

Figure 1.
A1-A3: The rodent photoneuroendocrine system and the circadian rhythm of melatonin. A1: BV: Blood vessels. ICN: Internal carotid nerves. IMC: Neurons of the intermediolateral columns of the spinal cord (SC). ipRGC: Intrinsically photosensitive retinal ganglion cells. MEL: Melatonin. NC: Nervi conarii. PG: Pineal gland. PVN: Paraventricular nuclei. RHT: Retinohypothalamic tracts. SCG: Superior cervical ganglia. SCN: Suprachiasmatic nuclei. A2: MEL-producing rat pinealocytes immunoreactive for serotonin (red; white arrowheads in the inset). White arrows: Interstitial cells negative for serotonin. DAPI: Nuclear marker 4',6-diamidino-2-phenylindole (blue). Immunofluorescence and confocal microscopy; objective: 10X, scale bar: 150 μm (inset: 60X with 4X zoom, scale bar: 15 μm.). See Ibañez Rodriguez et al. [9] for further details about animal procedures and immunolabeling protocols. A3: Rat superior cervical ganglion and surrounding anatomical landmarks. CCA: Common carotid artery. ECA: External carotid artery. ECN: External carotid nerve. HN: Hypoglossal nerve. ICA: Internal carotid artery. IJV: Internal jugular vein. ST: Sympathetic trunk. VN: Vagus nerve. Modified from Savastano et al. [10], where further details about animal procedures can be found. The reproduction of this copyrighted material was authorized by Elsevier. B: Melatonin biosynthetic pathway. AADC: Aromatic L-amino acid decarboxylase. AANAT: Arylalkylamine N-acetyltransferase. ASMT: Acetylserotonin O-methyltransferase. CSF: Cerebrospinal fluid. TPH: Tryptophan hydroxylase. III V: Third ventricle. C: Adrenergic regulation of melatonin synthesis at night. α1B-ADR: α1B adrenergic receptor. AC: Adenylate cyclase. ATP: Adenosine triphosphate. β1-ADR: β1 adrenergic receptor. cAMP: Cyclic adenosine monophosphate. CRE: cAMP responsive element. G: G proteins. mRNA: Messenger ribonucleic acid. pCREB: Phosphorylated form of cAMP responsive element-binding protein (CREB). PKA: Protein kinase A. PLC: Phospholipase C. PKC: Protein kinase C. D1-D2. Transcriptionally distinguished cell types and subtypes within the adult rat pineal gland. D1: Distribution of the cell types profiled in Mays et al. [11] during the light (L) and dark (D) phases of the L:D cycle. VLMC: Vascular and leptomeningeal cells. D2: Crosstalk between α- and β-pinealocytes to produce nocturnal melatonin in a coordinated and efficient manner. See Mays et al. [11] for further details.

2.1 Cellular composition of the mature pineal gland

Melatonin (MEL) is produced within the PG by its predominant cell population, the pinealocytes (Pc). One of the most modern classifications of rat Pc came with the application of single-cell RNA sequencing (scRNA-seq) (**Figure 1**) [11, 16]. This state-of-the-art technology provides gene expression profiles of isolated and individualized cells. Nowadays, it is accepted that at least two subtypes of pinealocytes, α-Pc and β-Pc, coexist and crosstalk in the rat PG, to produce MEL in a coordinated and efficient manner. β-pinealocytes are more abundant than α-Pc, but α-Pc are more effective in catalyzing the last step in the MEL biosynthetic pathway. The scRNA-seq analysis also discriminated interstitial cells. Among these transcriptionally distinguished non-pinealocyte cells are three astrocytes (α, β, and γ), two microglial subtypes (α and β), endothelial cells (EC), and vascular and leptomeningeal cells (VLMC). Several studies have shown that some non-pinealocyte cells modulate MEL production by pinealocytes, under both homeostatic and pathological conditions [17]. With respect to EC, they represent key elements within the PG because they form the inner lining of all blood vessels (BV) that make up its vast circulatory network, which are mainly fenestrated capillaries [8]. Therefore, the PG's blood vessels are more permeable and less selective than the tightly regulated blood-brain barrier (BBB) present in most of the central nervous system (CNS) [18]. The PG is included as one of the seven circumventricular organs (CVO) in the brain, and all of them have an incomplete barrier [8, 19]. This characteristic allows CVO to function as an intermediary pathway between the brain and the periphery, for bidirectional trafficking and interaction.

2.2 Melatonin synthesis by pinealocytes

Melatonin (MEL) is synthesized by pinealocytes (Pc) at night, via a multienzymatic pathway driven mainly by rhythmic neural inputs [2, 6, 7]. Circulating L-tryptophan (Trp) is an essential amino acid that acts as the biosynthetic precursor of the MEL molecule (**Figure 1**). Trp is hydroxylated and then decarboxylated enzymatically within the Pc cytoplasm. The product of these two reactions is serotonin or 5-hydroxytryptamine (5-HT). Serotonin is then converted into MEL by acetylation, followed by methylation [4]. The enzymes that catalyze the last two reactions, AANAT (Arylalkylamine N-acetyltransferase) and ASMT (Acetylserotonin O-methyltransferase), respectively, represent adjustable key points of the MEL-producing pathway [2, 11, 20]. Sympathetic axons stemming from neurons located in both superior cervical ganglia (SCG), provide the norepinephrine (NE) neurotransmitter signal that is the main regulator of MEL production. NE impacts the MEL biosynthetic machinery at different levels, from gene expression to enzyme activities, among other target points (**Figure 1**) [6, 7].

2.3 Sympathetic innervation of the mammalian pineal gland

The mammalian PG receives a wide range of afferent nerve fibers and, therefore, it can be influenced by a plethora of neurotransmitters [6, 8]. Efferent projections from the PG have also been described, but only for some species and at particular ontogenetic stages [8]. Among the afferent innervations, sympathetic axons, originating from both the right and left SCG, are a fundamental regulatory element of PG rhythmicity in mammals (**Figure 2**) [2]. Classic transcriptomic and neurotranscriptomic studies have shown that essentially all aspects of PG biology are subject to neural control. These aspects include thousands of genes associated with either MEL-related or MEL-unrelated functions, such as immune/inflammatory response and thyroid hormone signaling [11, 22–24].

2.3.1 Superior cervical ganglia

The SCG are the uppermost ganglia of the paravertebral sympathetic chain. They are well-defined structures with a variable number of neurons, which receive inputs from preganglionic fibers ascending in the sympathetic trunk (ST) (**Figure 1**) [10, 25, 26]. SCG neurons, mainly via the external and internal carotid nerves (ECN and ICN), establish a wide field of synapsis in the neck, face, and intracranial areas. The SCG not only innervate the pineal gland (PG), but also the hypophysis and median eminence, the thyroid and parathyroid glands, and the Muller's muscles (MM) that control the position of the upper eyelids (palpebral position). An important distinction is that the PG and the MM are innervated differently. Nerve fibers from both the right and left ICN innervate the PG bilaterally. Whereas for the MM, each MM is innervated unilaterally via efferent sympathetic axons present in the homolateral ICN. This innervation difference is used to evaluate the success of the SCG-related surgical procedures that are discussed in Section 2.3.3 (**Figure 2**). For all SCG targets and under tissue-specific stimuli (e.g., lights off for the PG), SCG-derived nerve terminals mainly release norepinephrine (NE) into the synaptic cleft and into the perivascular space. Additionally, other neuropeptides, such as the neuropeptide Y (NPY) in the PG, have been identified as sympathetic co-neurotransmitters [8]. The concentration of NE in the synaptic (or synaptic-like) gaps is affected by simple diffusion and uptake rates. NE uptake includes both its transport back into presynaptic nerve ends and its recapture by neighboring cells. NE re-uptake by sympathetic nerve terminals is crucial for stimulus termination, and for removal and deactivation of circulating stress-induced catecholamines. NE passes the message to the targets by stimulating specific adrenergic receptors on their cell membranes.

2.3.2 Adrenergic reception in the mature pineal gland

Pinealocytes (Pc) are the MEL-producing cells within the PG. As mentioned, Pc express adrenergic receptors on their cell membranes. These adrenoceptors bind and respond to the nocturnal NE released into the perivascular space from the sympathetic nerve ends (**Figure 1**). A recent scRNA-seq study confirmed the expression of two catecholamine receptor genes, *Adrb1* and *Adra1b*, in both α-Pc and β-Pc in the rat PG [11]. These genes encode β1 and α1B adrenergic receptors, respectively. Additionally, low levels of both transcripts were found in all the non-pinealocyte cells as well, with the exception that none were found in β-microglial cells. β1-ADR and α1B-ADR are seven-transmembrane (7TM) domain receptors that belong to the G protein-coupled receptor (GPCR) superfamily. The NE activation of these adrenoceptors triggers cooperative signaling pathways and several second messengers (e.g.: cyclic adenosine monophosphate, cAMP, and Ca^{2+}) that impact the whole Pc, including its nucleus and transcriptome (e.g., NE induces the expression of the *Aanat* gene, which encodes the enzyme AANAT, via the phosphorylated form of cAMP responsive element-binding protein, pCREB) [6, 7, 27–29]. As soon as *de novo* MEL is synthesized, it is released immediately into the bloodstream and into the cerebrospinal fluid (CSF) [3, 30]. MEL is produced during the dark phase of the L:D cycle and it is used

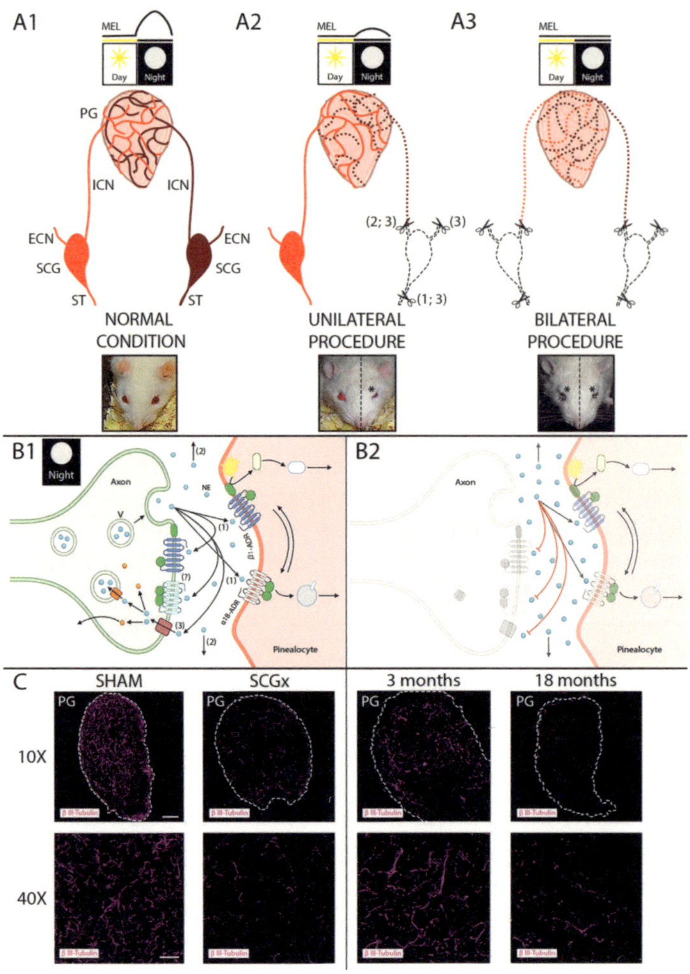

Figure 2

A1-A3: Surgical procedures related to the superior cervical ganglia and their impacts on pineal rhythmicity and palpebral position. A1: Bilateral sympathetic innervation of the pineal gland (PG) by both the right and left internal carotid nerves (ICN). This innervation provides nocturnal norepinephrine (NE), which drives the rhythmic synthesis of the hormone melatonin (MEL). ECN: External carotid nerves. SCG: Superior cervical ganglia. ST: Sympathetic trunks. A2: Reduction in the circulating MEL levels and homolateral blepharoptosis (*) after unilateral disruption of the sympathetic innervation. (1) Decentralization, by removal of a segment of the afferent sympathetic trunk (STx; lesion of preganglionic axons with undamaged ganglion in situ). (2) Denervation, by removal of a portion of the ICN (ICNx; lesion of postganglionic axons with undamaged ganglion in situ). (3) Ganglionectomy, by complete excision of the ganglion (SCGx; ablation of neuronal cell bodies). A3: Disappearance of the MEL circadian rhythm after a bilateral procedure. Surgery efficiency can be confirmed by observing palpebral ptosis in both eyes (*). The rat image also shows signs of chromodacryorrhoea (red tears). B1-B2: Norepinephrine re-uptake by nerve ends is abolished in the ganglionectomy model. B1: Norepinephrine (NE; light blue circles) released in the synaptic-like gaps, binds to specific adrenergic receptors (1) and diffuses outside the cleft (2). In addition, NE is transported back into the presynaptic nerve terminals in the healthy PG (3). NE re-uptake is crucial for stimulus termination, and for removal and deactivation of circulating stress-induced catecholamines (Orange circles: NE metabolites). α1B-ADR: α1B adrenergic receptor. β1-ADR: β1 adrenergic receptor. V: Vesicles. (?) Unknown. B2: NE re-uptake is abolished in the degenerating nerve terminals after SCGx. This is a difference with the STx procedure. C: Degeneration of the sympathetic nerve fibers within the pineal gland with age and after bilateral superior cervical ganglionectomy. Sections of rat pineal glands were immunolabeled for β III-tubulin, a marker of nerve fibers. SCGx: Bilateral ganglionectomy. SHAM: Bilateral sham surgery. Pineal glands from young (3 months) and aged (18 months) rats are shown. Immunofluorescence and confocal microscopy; objective: 10X, scale bar: 150 μm; objective: 40X, scale bar: 50 μm. See Ibañez Rodriguez et al. [9, 21] and Savastano et al. [10] for further details about animal procedures and immunolabeling protocols. The reproduction of the copyrighted rat images was authorized by Elsevier.

to disseminate the nighttime circadian status to all cells of the body, via specific MT1 and MT2 MEL receptors on the target cell membranes [31]. Nocturnal MEL production subsides towards late night and is shutdown during the daytime, in response to both extra-Pc mechanisms (e.g.: NE diffusion and uptake) and intra-Pc mechanisms (e.g.: feedback inhibition) [2, 6, 7].

2.3.3 SCG-related surgical procedures

Different procedures have been extensively used to study the sympathetic innervation of the mammalian PG, including surgical and pharmacological interventions, and electric stimulation [10, 23, 26, 32, 33]. Right and left ICN ascend from each SCG via the internal carotid arteries (ICA) and extend further to form the *nervi conarii* (NC) (**Figure 1**). The NC penetrate the PG at its dorso-posterior border and then ramify throughout the whole organ. Sometimes, bundles of axons from the two NC become fused before entering the gland. The sympathetic nerve fibers that innervate the PG along its vasculature, arise from a small population of neurons (not all neurons) that are rostrally dispersed in each SCG. When SCG-related surgeries are performed to completely suppress PG rhythmicity, both SCG must be isolated and manipulated in order to shut down the neural NE source bilaterally (**Figure 2**). A complete and permanent disruption of the MEL rhythm is achieved only when the influence of both NC is fully and irreversibly disrupted. In circadian biology, SCG-related surgeries are preferred to intracranial ones, such as a suprachiasmatic nuclei lesion (SCNx), due to the technical complications and the wider physiological impacts associated with these more invasive neurosurgeries. Because of the well-defined anatomy of the SCG and the surrounding structures, three types of surgical procedures can be executed to influence PG rhythmicity: (1) decentralization, by removal of a segment of the afferent sympathetic trunk (STx; lesion of preganglionic axons with undamaged ganglion *in situ*); (2) denervation, by removal of a portion of the ICN (ICNx; lesion of postganglionic axons with undamaged ganglion *in situ*), and (3) ganglionectomy, by complete excision of the ganglion (SCGx; ablation of neuronal cell bodies) (**Figure 2**). For those researchers who are interested in incorporating these procedures as routine techniques, there are straightforward protocols available in the literature, which provide step-by-step descriptions, illustrated with amazing images and detailed videos [10, 32, 33]. Each SCG-related surgery has advantages and disadvantages. For example, after a latency period following SCGx, the innervated target is seen to change through phases before stabilizing: (1) sympathetic stimulation due to the Wallerian degeneration of the sympathetic nerve terminals and a supraliminal release of neurotransmitter in the first few days following surgery (acute SCGx), and (2) sympathetic deprivation after the first post-surgery week (chronic SCGx). One complication of the SCGx is that it cannot prevent the influence of circulating stress-released catecholamines on the target tissue, because the local NE re-uptake mechanism is abolished in the degenerating sympathetic nerve ends (**Figure 2**). This obliges the users to exhaustively control animal housing conditions to eliminate any kind of stressor during the whole experimental time, even during animal euthanasia. On the contrary, STx preserves presynaptic re-uptake and the capacity to remove and deactivate circulating catecholamines. Additionally, SCGx induces microgliosis with damaging consequences over the PG parenchyma [21]. For these reasons, researchers prefer STx over SCGx. As mentioned, in the case of the PG, both SCG must be successfully manipulated to abolish the MEL circadian rhythm completely and permanently. This can be confirmed in a calm animal by bilateral blepharoptosis (palpebral ptosis), a sign commonly considered to evaluate surgery effectiveness (**Figure 2**) [10]. This sign is used because both the PG and the MM are innervated by the ICN (see Section 2.3.1).

2.4 Ontogeny of the mammalian pineal gland and its relationship with the sympathetic innervation

The PG emerges as an evagination of the roof of the diencephalon, late in the embryonic (E) period (E14-E15 for rat) [8, 9]. The basis of pineal morphogenesis, and the dynamic and intricate network of transcription factors (TF) involved in the establishment and maintenance of the pineal phenotype, have been extensively characterized, as well as the consequences of certain gene mutations on these mechanisms [28, 34–39]. Cells that are positive for the essential ontogenetic TF Pax6 and the intermediate filament protein vimentin (Vim) are present in the pineal primordium. The $Pax6^{+}/Vim^{+}$ precursor cells divide and go through an intrinsically and spatially programmed transformation, giving rise to pinealoblasts, which then mature perinatally to become pinealocytes. Astrocytes also derive from the $Pax6^{+}/Vim^{+}$ precursors, but later than Pc. Beyond the well-characterized sympathetic regulation of PG rhythmicity in mammals, researchers have questioned what role sympathetic innervation may have on the definition and fate of the pinealocyte lineage. Disruption of the SCG-derived innervation of the rat PG at 5 days after birth (P5; P: postnatal), by either STx or SCGx, did not substantially affect the establishment of the pineal-defining transcriptome (e.g., almost unaltered expression of the *Asmt* gene, which encodes the enzyme ASMT) [23]. As expected, both neonatal SCG-related surgeries did disrupt NE-dependent rhythms in the mature gland, including the circadian rhythm of melatonin (MEL). This suggests that functional sympathetic innervation might not be essential for pinealocyte definition, as it is for its circadian function. These results are consistent with previous classic reports about the ontogeny of adrenoceptors and the postnatal appearance of rhythms in adrenergic reception and signaling transduction, and in MEL-related enzymes [7, 40]. However, further comprehensive studies are necessary to confirm or not whether sympathetic and non-sympathetic innervations do indeed participate in the fine definition of the PG phenotype. This might include interventions earlier than P5, for example.

2.5 Sympathetic dysfunction

In general, an abnormal melatonin rhythm has been associated with a wide spectrum of human pathologies, including sleep disorders, obesity, diabetes, cancer, and genetic, trauma-induced, neurological, and neurodegenerative disorders [1, 4, 41–43]. Our modern life, with the use of artificial lighting, time-shifted work schedules, and travel jet lag, contributes to alterations in circulating MEL levels in humans [42, 44]. MEL production normally subsides as we age and is aggravated by the more prolonged life expectancy of current generations [45]. This directly affects sleep patterns and mental alertness, but it also has short-term and long-term impacts on overall health. Taking MEL supplements has become a popular remedial therapy when endogenous MEL production is deemed to be deficient or altered. However, basic questions regarding MEL consumption and optimal dosage have not yet been resolved. Nevertheless, MEL supplementation is being used mainly to improve sleep quality and to treat certain sleep disorders. Additionally, it is consumed to attenuate tissue damage due to the cytoprotective properties of MEL itself. Further studies are needed to clarify the cellular and molecular mechanisms behind the altered MEL patterns for each of these pathological landscapes. In addition, the therapeutical potential of MEL and its analogs for a wide range of human pathologies needs further investigation [4, 5, 43]. As mentioned, the experimental disruption of the SCG-derived innervation, when it is executed bilaterally and irreversibly, shuts down NE-dependent pineal rhythmicity. In humans, there are several pathological conditions that may be accompanied by primary or secondary sympathetic alterations, that therefore may cause pineal dysfunction. For example, spinal cord injuries (SCI) can be a primary mechanism for sympathetic abnormality [46]. SCI at the upper thoracic segments or higher (cervical injury) may sever the descending axons from the hypothalamic PVN, which connect the SCN to the preganglionic neurons located in the IMC of the spinal cord (SC). As a result, these IMC neurons may not establish functional synapsis with SCG ganglionic cells, which would impair or abolish nocturnal MEL synthesis [47]. In fact, patients with certain upper SCI sometimes experience altered levels of circulating MEL, and disrupted sleep patterns and behaviors. Exogenous MEL has been used to ameliorate SCI consequences due to both MEL's chronobiotic and cytoprotective qualities [48]. On the other hand, a large number of studies has pointed to an age-related loss of the pineal function in both animals and humans (e.g.: elderly individuals, and preclinical and clinical patients with aging-related pathologies, such as Alzheimer's disease and Parkinson's disease) [4, 43, 45, 49–51]. This loss has been linked to some or all the following features: anatomical abnormalities, reduced number of pinealocytes and variable numbers of glial cells, fibrosis, calcification, inflammation, altered CG expression and clock functionality, disconnection from the master circadian clock at the hypothalamic SCN, and impaired sympathetic regulation [49–54]. Sympathetic dysregulation may involve the loss of nerve terminals, as well as age-induced neuroaxonal dystrophy (NAD) of distal axons, altered denervation supersensitivity, and a decrease in adrenoceptor reception and responsiveness, among other mechanisms (**Figure 2**) [53, 55]. The sequence and progression of these structural and mechanistic alterations during aging and aging-associated pathologies have not yet been fully clarified. Acute and chronic MEL supplementation, in both early and late stages of these conditions, also warrants further investigation. The use of exogenous MEL supplements represents, however, a promising remedial therapy especially for those patients in preclinical phases due to both its chronobiotic and its cytoprotective properties [51, 56, 57].

3. Conclusions

In mammals, sympathetic innervation plays a key regulatory role in pineal biology and its circadian production of the hormone melatonin. A proper melatonin rhythm, with its classic level rise at night, requires an intact photoneuroendocrine system, that transduces light information from the external environment into the hormonal cue. To do this, a hierarchical series of oscillators, that includes the brain's master circadian clock, operates in a coordinated manner to assure that this information is communicated to the pineal gland through the right and left superior cervical ganglia. The well-characterized structural and functional features of this association have made the pineal gland one of the preferred models for understanding the role of sympathetic innervation in health and disease. A wide spectrum of human pathologies may be accompanied by pineal dysfunction. This may be related to different forms of sympathetic abnormalities. Further studies are necessary to delve deeper into the cellular and molecular mechanisms responsible for altered melatonin rhythms in these pathological landscapes. In addition, further efforts are needed to elucidate whether melatonin supplementation is useful to prevent or to ameliorate the impacts of these conditions on health and life quality.

Acknowledgements

We thank Raymond D. Astrue for editing the manuscript. This work was supported by grants from CONICET (Argentina; PUE 2017; http://www.conicet.gov.ar), and ANPCyT (Argentina; PICT 2017-499 and PICT 2021-314; http://www.agencia.mincyt.gob.ar).

Conflict of interest

The authors declare no conflict of interest.

Author details

Martin Avila[1†], Carlos L. Freites[1†], Elena Vásquez[1], Juan B. Amiotti[1], Janina Borgonovo[2] and Estela M. Muñoz[1*]

1 Institute of Histology and Embryology of Mendoza (IHEM), National University of Cuyo (UNCuyo), National Scientific and Technical Research Council (CONICET), Mendoza, Argentina

2 Laboratory of Experimental Ontogeny (LEO), Faculty of Medicine, Institute of Biomedical Sciences, Universidad de Chile, Santiago, Chile

*Address all correspondence to: munoz.estela@fcm.uncu.edu.ar; emunoz@conicet-mendoza.gob.ar

†These authors contributed equally to this chapter.

References

[1] Pandi-Perumal SR, Srinivasan V, Maestroni GJ, Cardinali DP, Poeggeler B, Hardeland R. Melatonin: Nature's most versatile biological signal? The FEBS Journal. 2006;**273**(13):2813-2838

[2] Borjigin J, Zhang LS, Calinescu AA. Circadian regulation of pineal gland rhythmicity. Molecular and Cellular Endocrinology. 2012;**349**(1):13-19

[3] Tan DX, Manchester LC, Reiter RJ. CSF generation by pineal gland results in a robust melatonin circadian rhythm in the third ventricle as an unique light/dark signal. Medical Hypotheses. 2016;**86**:3-9

[4] Arendt J, Aulinas A. Physiology of the pineal gland and melatonin. In: Feingold KR, Anawalt B, Blackman MR, Boyce A, Chrousos G, Corpas E, et al., editors. Endotext. South Dartmouth (MA), MDText.com, Inc. 2000; 2022

[5] Givler D, Givler A, Luther PM, Wenger DM, Ahmadzadeh S, Shekoohi S, et al. Chronic Administration of Melatonin: Physiological and clinical considerations. Neurology International. 2023;**15**(1):518-533

[6] Simonneaux V, Ribelayga C. Generation of the melatonin endocrine message in mammals: A review of the complex regulation of melatonin synthesis by norepinephrine, peptides, and other pineal transmitters. Pharmacological Reviews. 2003;**55**(2):325-395

[7] Maronde E, Stehle JH. The mammalian pineal gland: Known facts, unknown facets. Trends in Endocrinology and Metabolism. 2007;**18**(4):142-149

[8] Møller M, Baeres FM. The anatomy and innervation of the mammalian pineal gland. Cell and Tissue Research. 2002;**309**(1):139-150

[9] Ibañez Rodriguez MP, Noctor SC, Muñoz EM. Cellular basis of pineal gland development: Emerging role of microglia as phenotype regulator. PLoS One. 2016;**11**(11):e0167063

[10] Savastano LE, Castro AE, Fitt MR, Rath MF, Romeo HE, Muñoz EM. A standardized surgical technique for rat superior cervical ganglionectomy. Journal of Neuroscience Methods. 2010;**192**(1):22-33

[11] Mays JC, Kelly MC, Coon SL, Holtzclaw L, Rath MF, Kelley MW, et al. Single-cell RNA sequencing of the mammalian pineal gland identifies two pinealocyte subtypes and cell type-specific daily patterns of gene expression. PLoS One. 2018;**13**(10):e0205883

[12] Patton AP, Hastings MH. The mammalian circadian time-keeping system. The Journal of Huntington's Disease. Vol. Pre-press, no. Pre-press, 2023. pp. 1-14

[13] Muñoz E, Brewer M, Baler R. Circadian Transcription. Thinking outside the E-box. The Journal of Biological Chemistry 2002;277(39):36009-36017.

[14] Muñoz E, Baler R. The circadian E-box: When perfect is not good enough. Chronobiology International. 2003;**20**(3):371-388

[15] Muñoz E, Brewer M, Baler R. Modulation of BMAL/CLOCK/E-box complex activity by a CT-rich cis-acting element. Molecular and Cellular Endocrinology. 2006;**252**(1-2):74-81

[16] Ding X, Pan T, Tian Q, Huang W, Hayashi LS, Liu Q, et al. Profiling temporal changes of the pineal transcriptomes at single cell level upon neonatal HIBD. Frontiers in Cell and Development Biology. 2022;**10**:794012

[17] da Silveira C-MS, Pinato L, Tamura EK, Carvalho-Sousa CE, Markus RP. Glia-pinealocyte network: The paracrine modulation of melatonin synthesis by tumor necrosis factor (TNF). PLoS One. 2012;**7**(7):e40142

[18] Benz F, Liebner S. Structure and function of the blood-brain barrier (BBB). Handbook of Experimental Pharmacology. 2022;**273**:3-31

[19] Muñoz EM. Microglia in circumventricular organs: The pineal gland example. ASN Neuro. 2022;**14**:17590914221135697

[20] Klein DC. Arylalkylamine N-acetyltransferase: "the Timezyme". The Journal of Biological Chemistry. 2007;**282**(7):4233-4237

[21] Ibañez Rodriguez MP, Galiana MD, Rasmussen JA, Freites CL, Noctor SC, Muñoz EM. Differential response of pineal microglia to surgical versus pharmacological stimuli. The Journal of Comparative Neurology. 2018;**526**(15):2462-2481

[22] Bailey MJ, Coon SL, Carter DA, Humphries A, Kim JS, Shi Q, et al. Night/day changes in pineal expression of >600 genes: Central role of adrenergic/cAMP signaling. The Journal of Biological Chemistry. 2009;**284**(12):7606-7622

[23] Hartley SW, Coon SL, Savastano LE, Mullikin JC, Program NCS, Fu C, et al. Neurotranscriptomics: The effects of neonatal stimulus deprivation on the rat pineal transcriptome. PLoS One. 2015;**10**(9):e0137548

[24] Chang E, Fu C, Coon SL, Alon S, Bozinoski M, Breymaier M, et al. Resource: A multi-species multi-timepoint transcriptome database and webpage for the pineal gland and retina. Journal of Pineal Research. 2020;**69**(3):e12673

[25] Ladd FV, Ladd AA, da Silva AA, Coppi AA. Stereological and allometric studies on neurons and axo-dendritic synapses in superior cervical ganglia. International Review of Cell and Molecular Biology. 2014;**311**:123-155

[26] Lumsden SC, Clarkson AN, Cakmak YO. Neuromodulation of the pineal gland via electrical stimulation of its sympathetic innervation pathway. Frontiers in Neuroscience. 2020;**14**:264

[27] Baler R, Covington S, Klein DC. The rat arylalkylamine N-acetyltransferase gene promoter. cAMP activation via a cAMP-responsive element-CCAAT complex. The Journal of Biological Chemistry. 1997;**272**(11):6979-6985

[28] Farias Altamirano LE, Freites CL, Vasquez E, Muñoz EM. Signaling within the pineal gland: A parallelism with the central nervous system. Seminars in Cell & Developmental Biology. 2019;**95**:151-159

[29] Farias Altamirano LE, Vásquez E, Freites CL, Ibañez JE, Guido ME, Muñoz EM. Spatio-temporal dynamics of nuclear CREB1: What does it mean? bioRxiv. 2022. DOI: 10.1101/2022.06.26.497665

[30] Yu H, Dickson EJ, Jung SR, Koh DS, Hille B. High membrane permeability for melatonin. The Journal of General Physiology. 2016;**147**(1):63-76

[31] Jockers R, Delagrange P, Dubocovich ML, Markus RP, Renault N, Tosini G, et al. Update on melatonin

receptors: IUPHAR review 20. British Journal of Pharmacology. 2016;**173**(18):2702-2725

[32] Madhani SI, Klein DC, Muñoz EM, Savastano LE. Surgical techniques and nuances for superior cervical Ganglionectomy and decentralization in rats. Methods in Molecular Biology. 2022;**2550**:53-62

[33] Wang Q, Chen CH, Xu H, Deborde S, Wong RJ. Surgical technique for superior cervical Ganglionectomy in a murine model. Journal of Visualized Experiments. 2022;**190**. DOI: 10.3791/64527

[34] Muñoz EM, Bailey MJ, Rath MF, Shi Q, Morin F, Coon SL, et al. NeuroD1: Developmental expression and regulated genes in the rodent pineal gland. Journal of Neurochemistry. 2007;**102**(3):887-899

[35] Ochocinska MJ, Muñoz EM, Veleri S, Weller JL, Coon SL, Pozdeyev N, et al. NeuroD1 is required for survival of photoreceptors but not pinealocytes: Results from targeted gene deletion studies. Journal of Neurochemistry. 2012;**123**(1):44-59

[36] Rath MF, Rohde K, Klein DC, Møller M. Homeobox genes in the rodent pineal gland: Roles in development and phenotype maintenance. Neurochemical Research. 2013;**38**(6):1100-1112

[37] Castro AE, Benitez SG, Farias Altamirano LE, Savastano LE, Patterson SI, Muñoz EM. Expression and cellular localization of the transcription factor NeuroD1 in the developing and adult rat pineal gland. Journal of Pineal Research. 2015;**58**(4):439-451

[38] Yamazaki F, Møller M, Fu C, Clokie SJ, Zykovich A, Coon SL, et al. The Lhx9 homeobox gene controls pineal gland development and prevents postnatal hydrocephalus. Brain Structure & Function. 2015;**220**(3):1497-1509

[39] Muñoz EM. Microglia-precursor cell interactions in health and in pathology. Biocell. 2018;**42**(2):41-45

[40] Cantor EH, Clark MB, Weiss B. Effect of sympathetic input on ontogeny of beta-adrenergic receptors in rat pineal gland. Brain Research Bulletin. 1981;**7**(3):243-247

[41] Woodford EC, McLay L, France KG, Blampied NM, Gibbs R, Swan CE, et al. Endogenous melatonin and sleep in individuals with rare genetic neurodevelopmental disorders (RGND): A systematic review. Sleep Medicine Reviews. 2021;**57**:101433

[42] Muscogiuri G, Poggiogalle E, Barrea L, Tarsitano MG, Garifalos F, Liccardi A, et al. Exposure to artificial light at night: A common link for obesity and cancer? European Journal of Cancer. 2022;**173**:263-275

[43] Ahmad SB, Ali A, Bilal M, Rashid SM, Wani AB, Bhat RR, et al. Melatonin and health: Insights of melatonin action, biological functions, and associated disorders. Cellular and Molecular Neurobiology. 2023;**43**:2437-2458

[44] Zielinska-Dabkowska KM, Schernhammer ES, Hanifin JP, Brainard GC. Reducing nighttime light exposure in the urban environment to benefit human health and society. Science. 2023;**380**(6650):1130-1135

[45] Godfrey S, Iversen HK, West AS. Melatonin profile in healthy, elderly subjects - a systematic literature review. Chronobiology International. 2022;**39**(4):476-492

[46] Wulf MJ, Tom VJ. Consequences of spinal cord injury on the sympathetic nervous system. Frontiers in Cellular Neuroscience. 2023;**17**:999253

[47] Whelan A, Halpine M, Christie SD, McVeigh SA. Systematic review of melatonin levels in individuals with complete cervical spinal cord injury. The Journal of Spinal Cord Medicine. 2020;**43**(5):565-578

[48] Zhang Y, Zhang WX, Zhang YJ, Liu YD, Liu ZJ, Wu QC, et al. Melatonin for the treatment of spinal cord injury. Neural Regeneration Research. 2018;**13**(10):1685-1692

[49] Wu YH, Swaab DF. The human pineal gland and melatonin in aging and Alzheimer's disease. Journal of Pineal Research. 2005;**38**(3):145-152

[50] Breen DP, Nombela C, Vuono R, Jones PS, Fisher K, Burn DJ, et al. Hypothalamic volume loss is associated with reduced melatonin output in Parkinson's disease. Movement Disorders. 2016;**31**(7):1062-1066

[51] Asadpoordezaki Z, Coogan AN, Henley BM. Chronobiology of Parkinson's disease: Past, present and future. The European Journal of Neuroscience. 2023;**57**(1):178-200

[52] Reuss S, Spies C, Schroder H, Vollrath L. The aged pineal gland: Reduction in pinealocyte number and adrenergic innervation in male rats. Experimental Gerontology. 1990;**25**(2):183-188

[53] Schmidt RE, Dorsey DA, Parvin CA, Beaudet LN. Sympathetic neuroaxonal dystrophy in the aged rat pineal gland. Neurobiology of Aging. 2006;**27**(10):1514-1523

[54] Wu YH, Fischer DF, Kalsbeek A, Garidou-Boof ML, van der Vliet J, van Heijningen C, et al. Pineal clock gene oscillation is disturbed in Alzheimer's disease, due to functional disconnection from the "master clock". The FASEB Journal. 2006;**20**(11):1874-1876

[55] Weiss B, Greenberg LH, Cantor E. Denervation supersensitivity and beta-adrenergic receptors as a function of age. Advances in Biochemical Psychopharmacology. 1980;**21**:461-472

[56] Wu YH, Swaab DF. Disturbance and strategies for reactivation of the circadian rhythm system in aging and Alzheimer's disease. Sleep Medicine. 2007;**8**(6):623-636

[57] Smilowska K, van Wamelen DJ, Bloem BR. The multimodal effect of circadian interventions in Parkinson's disease: A narrative review. Parkinsonism & Related Disorders. 2023;**110**:105309

Chapter 6

The Brain-Like Enteric Nervous System

Flower M.J. Caycho Salazar, Deissy Herrera-Covarrubias, Genaro A. Coria-Ávila, Luis I. García-Hernández, María Rebeca Toledo-Cárdenas, María Elena Hernández-Aguilar and Jorge Manzo

Abstract

Understanding the autonomic supply at the gastrointestinal tract is one of the significant challenges for science. Its complex network of neurons exists on a broad evolutionary scale, from Hydra to mammals, and in a higher number than those found in the vertebrate spinal cord. Inside the gastrointestinal tract, enteric neurons regulate several functions with intrinsic processes and communicate with the other complex known as the microbiome. Outside the gastrointestinal tract, the enteric neurons project to the brain stem and spinal cord via the gut–brain axis. Furthermore, this enteric system has close functional relationships with the immune system for a rapid response to unhealthy food. The present chapter focuses on the structure, function, and pathologies of the enteric nervous system.

Keywords: enteric nervous system, gastrointestinal, microbiome, brain–gut axis, second brain

1. Introduction

The enteric nervous system (ENS) is a complex network of neurons and glia that regulates the physiology of the gastrointestinal tract. It is the largest division of the autonomic nervous system and is responsible for controlling several functions, including motility, secretion, blood flow, and immune surveillance. The ENS spans the entire length of the gastrointestinal tract and comprises over 100 million neurons in humans, a higher number than those found in the spinal cord. It is considered a second brain because it carries out specific functions that do not depend on the central nervous system (CNS). From an evolutionary perspective, however, the ENS could be considered the first brain, as it evolved early in development in multicellular organisms to allow for efficient food processing and digestion. Although such a discussion is out of the scope of this chapter, it is worth saying that nutrients imposed an evolutionary pressure on all living species because it is a critical vital need; in animals, it was necessary to adapt a precise autonomic control, leading to the rise of a primitive system that became the ENS in contemporary species. Undoubtedly, this later fact reveals that the ENS is the oldest region of the nervous system; the Hydra, for example, a 500 million years cnidarian [1], is the oldest known group with sensory neurons in the oral region to regulate feeding, and with clustered ganglion neurons at the hypostome-tentacle junction to trigger contraction burst pulses of the epithelium to allow movement and ingestion [2]. The organization of these neurons is of great significance since it persisted throughout evolution and is observed from Hydras up to humans [3].

The complex circuitry of the ENS allows for the organization of both local and long-distance reflexes. These reflexes start with sensory neurons that detect changes in the gut's environment, such as the presence of food, and then relay this information to the motor neurons that control gut function. At a glance, the ENS is an essential component of the autonomic nervous system, and its complex and sophisticated functions make it a critical player in regulating gastrointestinal function. Furthermore, its ability to function independently of the CNS, coupled with its communication with the brain, highlights the importance of this intricate neural network in maintaining overall health and well-being.

The autonomic control of the intestine was first described in 1847 by Robert Remak, who settled the basis for further descriptions by Georg Meissner (1852) and Leopold Auerbach (1862), establishing the initial studies about the ENS [4]. Now, data show that neurons and glia of the ENS have their embryological origin at the neural crest, and before and after arriving at the gastrointestinal tract, they differentiate into glia and different types of neurons [5]. A detailed analysis of cellular and molecular processes of ENS development is in an excellent review [6]. In brief, this system originates from neural crest cells (NCC) derived from the ectoderm at the neural tube. NCC delaminate, and during this epithelial-mesenchymal transition, other levels or axes of the neural tube arise, named cranial, cardiac, vagal, truncal, and sacral. Subsequently, they proliferate and migrate until colonize specific sites. In such a process, they differentiate into various types of cells to structure the tissues to make up the gastrointestinal tract [6, 7]. These tissues can be diverse: connective tissue, endocrine cells, glia, and enteric neurons. The process of cell differentiation is necessary for a functional ENS; hence it is gradual because markers for neuronal types appear and may continue to a particular postnatal stage [8]. Studies are increasing to determine which molecules are involved in cell differentiation, such as the transcription factors and signaling pathways. Many of these molecules are for neuronal differentiation, as the SRY-like high-mobility group (HMG)-box (Sox) family, Sox6, Sox10, Mash1 (now called Ascl1), Hand 2; and those known for glial differentiation, as the GDNF, Neurturin, and the signaling pathway Ret – Rearranged during transfection – Notch [9].

2. Neurons and glia

The neurons form at least two ganglionic nerve plexuses running along the submucose layer of the gastrointestinal tract, the inner and the outer submucose plexus, and even a third plexus observed in humans [10]. Also, they are called the myenteric or Auerbach's plexus, the submucous or Meissner's plexus, and the mucous plexus [11]. The first is a plexus that runs from the esophagus to the rectum, while the others are located mainly in the intestines, with some functions independent of the influence of the central nervous system [12]. Thus, the ENS is a specialized system with significant self-supporting processes.

There are different kinds of neurons in the ENS. In the 19th century, the Russian neuroscientist Alexandre S. Dogiel described three types of neurons at the ENS for the first time. The current terminology recognizes them as Type I (one axon and short dendrites), Type II (one axon and long dendrites), and Type III neurons (one axon and long tapering and branching dendrites found in the guinea pig); all of them also recognized as multipolar neurons [13]. Then, Type IV was described by Stach in the 1980s as a radiate multidendritic uniaxonal neuron with branches between the myenteric and submucous plexus [14], and Type V referring neurons with long dendrites observed in pigs and humans [15, 16].

The Dogiel Type II neurons in the guinea pig are primary afferent neurons [17] also found in humans in the stomach, small intestine, and colon [13]. The Dogiel Type I neurons show specific subdivisions depending on the shape; stubby neurons have short and stubby dendrites, spiny neurons with short and thorny dendrites, and hairy neurons with short and thin dendrites [13].

ENS glial cells outnumber neurons, as occurs in the central nervous system. They are flat and stellate-shaped, extended over neurons and neuronal processes, with a similar arrangement between vertebrate species [18]. The structure and molecular characteristics of enteric glia suggest that they are astrocytes-like cells [19] subdivided into Type I or protoplasmic glia, and Type II or fibrous glia, an organization determined by the microenvironment [20]. Also, there is a description of a Type III glia showing long and branched processes, and the Type IV referring to that glia on nerve fibers in the muscle layer; notwithstanding, a new proposal is to name them according to their location, cells in the myenteric and submucosal plexuses referred as EG_{MP} and EG_{SMP}, and cells in the mucosa and musculature as EG_{Mucosa} and EG_{IM} [21].

Although the classification of ENS neurons and glia is remarkable, as far as innovative methods become available, identifying and characterizing the structure of these cells will still be refined. The task will improve our understanding of gastrointestinal functions, diseases, and treatments.

3. Microbiome

The gastrointestinal microbiome (GM), previously known as the intestinal flora, is the world of microorganisms that live in the gut to support digestion and significantly impact health. It includes different bacterial taxonomic groups and their interrelations [22]; for example, 400+ bacterial species live just in the human colon [23]. No matter whether microbiomes also exist in the skin, mouth, or reproductive tract, GM is the most well-studied in humans. Thus, it is known that starting at birth, the GM is acquired from the mother during delivery and breastfeeding [24]. GM plays a significant role in breaking down food, extracting nutrients, and producing vitamins necessary for human health; consequently, it is modified during life depending on the environment of the subject [25]. Considering the complex ecosystem that several microbes species establish in the gut, the role they play for health, their physiology as a group that impacts health if modified, and the many functions that even affect behavior, the GM is considered an organ by itself [22, 26]. Such a statement is further consolidated due to the relationship between GM and ENS.

The afferent starting point is at the enteroendocrine cells within the gut. They are specialized cells that respond to ingested substances. Although not entirely known, they release hormones such as cholecystokinin to activate nerve pathways to the central nervous system [27]. Also, neurotransmitters such as the gamma-aminobutyric acid (GABA) play a specific role in activating the afferent pathways [28]. Whatever the milieu chemicals, the activation of enteroendocrine cells, in turn, activates the two main types of ENS afferent neurons, the extrinsic and intrinsic afferents, which differ depending on the location of their cell bodies, outside (extrinsic) or inside (intrinsic) the ENS [29]. Notwithstanding, the exact mechanisms by which the GM activates the ENS are yet not completely understood, but it represents the first interface between intestinal content and ENS to activate the afferent pathways of the brain–gut axis [30].

4. Brain–gut axis

Gastrointestinal afferent neurons are Dogiel Type II cells representing about 20% of ENS neurons. The intrinsic complex is formed by the intrinsic primary afferent neurons (IPAN), interneurons, and motor neurons, which organize local circuits within the ENS to trigger CNS-independent reflexes that regulate several aspects of gut function [31]. The extrinsic neurons have cell bodies in the dorsal root ganglia and the complex jugular-nodose ganglia at the jugular foramen. Extrinsic fibers from the stomach and upper intestine run from the gut to the CNS via the vagus and splanchnic nerves, and those from the distal intestine run via pelvic nerves [32].

The jugular ganglion is the smallest afferent cluster of sensory neurons of the vagus nerve [33] and also has neurons with similar properties as small dorsal root ganglion neurons, suggesting a nociceptive role [34]. The afferents project to the brain stem, specifically to the nucleus of the solitary tract (NTS), area postrema, and the upper cervical dorsal horn [35]. The nodose ganglion and its neurons are organized in a viscerotopic position, i.e., located inside the ganglion, depending on the origin of the afferent information. Then, they project central fibers through the solitary tract that synapse on neurons of the NTS located in the medulla [36]. NTS is a complex nucleus with projections to different cortex areas and brain nuclei, such as the insular cortex, frontal cortex, or thalamus [37]. It is a region for inputs from several regions, such as the insular cortex, paraventricular nucleus, hypothalamus, and amygdala [38].

The cell bodies of splanchnic afferents neurons are in the dorsal roots ganglia of the thoracolumbar spinal cord. Such neurons activate ascendent fibers in the spinothalamic, spinoreticular, and dorsal column pathways that carry information about noxious stimuli to different parts of the CNS, where it is interpreted as pain or discomfort [39]. However, the information from splanchnic afferents has a less graded sensation in response to distention, implying more intense or unpleasant sensations than vagal and pelvic afferents [40].

Pelvic afferent neurons are the third input pathway critical in sending information from the gastrointestinal tract to the CNS. These neurons are subdivided into two types based on their firing pattern: tonic and phasic. Tonic afferents become active by colonic distention and mainly consist of unmyelinated C fibers, while phasic afferents discharge at the onset and cessation of distention and include myelinated A-delta fibers [41]. They enter the spinal cord through the lumbosacral dorsal root ganglia and activate different ascending tracts [42].

Activation of afferent pathways triggers the different reflexes of the gastrointestinal system. Intrinsic activity is represented by ascending and descending reflexes to increase the luminal content and initiate peristalsis [29]. Ascending reflexes are excitatory pathways that induce the peristaltic contraction of circular muscles, and descending reflexes were described as inhibitory [43]. However, specific inhibitory and excitatory neurons exist in ascending and descending pathways [44]. These reflexes initiate following the enteroendocrine cell's activation of the IPANs, then activate interneuron and motor neurons to produce the appropriate response [45]. IPANs are located in submucosal or myenteric areas, and they respectively trigger peristaltic and secretory reflexes or stretch contraction reflexes via cholinergic pathways [46].

The extrinsic activity allows reflexes to perform tasks involving neurons in the CNS. Neurons at the dorsal motor nucleus of the vagus and at the NTS activate efferent pathways via the vagal outflow to the ENS [37]. Such pathways allow a fine modulation of gastrointestinal functions, mainly in the upper gastrointestinal tract, although the vago-vagal reflexes also include esophagogastric, gastrograstric, and duodenogastric reflexes that still need more studies [47]. It is noteworthy that vagal reflexes are not a fixed response, as observed in spinal reflexes. Instead, they are modulated, and the response depends on the demand of the gastrointestinal tract [48].

5. Immune system

A healthy gastrointestinal system depends on the immune system, notwithstanding it also depends on the collaborative work that immunity maintains with the microbiota and the ENS. The gastrointestinal tract is the region with more concentration of immune cells, mainly macrophages. The microbiota stimulates both macrophages and ENS neurons to synthesize and release, respectively, the bone morphogenic protein 2 (BMP2) and the colony stimulator factor 1 (CSF1). ENS neurons have receptors for BMP2 and macrophages for CSF1, representing the complex crosstalk signal circuit that exists to control gut function [49, 50].

Macrophages represent a diverse group of guard cells for the custody of the surrounding environment aimed to prevent infections [51]. Those in the smooth muscle of the gastrointestinal tract are in close contact with ENS neurons and regulate synaptic functions that include control of neuropeptides and neurotransmitters, but also receive activation from the ENS for the neuroimmune responses [52]. The comparison between lamina propia macrophages located in the epithelium close to the lumen, and those muscularis macrophages, show that they have particular responses to support the specialized interaction of the ENS and the immune system [53]. In the event of a response, such as inflammation, both macrophages and the ENS become active to restore homeostasis. Such activation is observed in local circuits but also in central ones as the inflammatory reflex, in which central neurons in the NTS and motor neurons of the dorsal nucleus of the vagus nerve become active [54].

6. Diseases

Several disorders are linked to dysfunctions of the ENS. For example, the so-called enteric neuropathies arise from the loss, degeneration, or functional impairment of enteric neurons, which may be congenital disabilities during development induced by infectious agents or conditions such as diabetes and neurodegenerative diseases [55]. Furthermore, specific dysfunctions or damage to the submucosal plexus are linked to gastrointestinal disorders and other disorders.

6.1 Irritable bowel syndrome (IBS)

According to the Rome IV criteria, the symptoms of IBS are frequent abdominal pain associated with bloating or the rhythm of evacuations, such as constipation, diarrhea, or both. Also, IBS is subdivided according to the defecation pattern, those with diarrhea IBS-D, constipation IBS-C, a mixed subtype IBS-M, and even those not yet subtyped, known as IBS-U [56]. Diagnosis includes the frequency criterion, i.e., if the abdominal pain occurs once a week, for at least 3 months, and the onset of symptoms with a minimum of six months before diagnosis [57]. IBS is recognized by altered gastrointestinal motility, characterized by accelerated GI transit in response to enteric ganglionitis in severe cases, carbohydrate malabsorption, bacterial overpopulation [58], visceral hypersensitivity, mucosal permeability, and altered microbiota [59].

The etiology of IBS has not yet been fully clarified, but there is sufficient evidence to link the immune system interaction with the ENS in the syndrome's pathophysiology. Notably, inflammation in IBS is marked and considered a significant feature in diagnosis, including in patients with postinfectious IBS [60]. The immune mast cells in the submucosal plexus trigger multiple inflammatory responses and generate a neuroimmune response when interacting with enteric neurons. The signaling of mast cells to enteric neurons is via neurotransmitters and neurohormones such as histamine and tryptase, which exert an excitatory function on the submucosal plexus, and an increased density of such cells is correlated with visceral hypersensitivity [61]. Genes are also involved, some supporting neuronal functions associated with IBS [56].

6.2 Hirschsprung disease (HSCR)

HSCR is a primary enteric neuropathy and one of the most frequent gastrointestinal motility disorders, showing the absence of enteric ganglia mainly in the colon. The congenital absence of ganglia neurons at the submucosal and myenteric plexuses occurs following a failure in the migration process of cells from the enteric neural crest to the hindgut; thus, this disease is known as a neurochristopathy but also is known as congenital megacolon or intestinal aganglionosis [62, 63]. The absence of ganglia produces a reduced or no peristalsis at all, causing intestinal occlusion because of the cessation of the expelling of fecal material. HSCR has an incidence of 1/5000 newborns and is more prevalent in males, in a 4:1 ratio [62].

6.3 Achalasia

Achalasia is a rare disorder affecting the motility of the esophageal region, characterized by the loss of enteric neurons and inhibitory postganglionic neurons that produces the absence of peristalsis of the tubular esophagus and impaired relaxation of the lower esophageal sphincter, involved in the swallow reflex. Symptoms include dysphagia, heartburn, regurgitation, chest pain, and weight loss. There is no total clarity of the etiology, but evidence exists that it is associated with autoimmune processes to still unknown antigens [64]. Furthermore, similar manifestations are found in some cases, as those related to the Chagas disease [65, 66].

6.4 Chronic constipation

Constipation by itself is not a disease, but it is considered a widespread gastrointestinal disorder that turns out to be the primary symptom to diagnose a disease.

During constipation, defecation is difficult, accompanied by pain and stiffness (Forootan 2018), which is more frequent in women [67]. It is classified into two types, primary and secondary. Primary refers to dysregulation of neuromuscular activity within the colon and rectum, sometimes called functional constipation, that includes irritable bowel syndrome, and slow-transit constipation, caused by dysfunction of the smooth muscle activity in the colonic region. The secondary is nonspecific because constipation can respond to multiple factors such as metabolic problems, intake of medications, diet, neurological disorders, or colon diseases [68].

6.5 Autism spectrum disorder (ASD)

ASD is a neurodevelopmental disorder characterized by two domains, social and communicative difficulties, and restricted and repetitive behaviors, that appear early in childhood. In addition to the whole manifestations, ASD is commonly accompanied by many comorbid conditions that include a significant prevalence of gastrointestinal alterations such as chronic gastrointestinal dysfunction (diarrhea, constipation, reflux, etc.), or food intolerance [69, 70]. Also, ASD children show physiological alterations such as increased intestinal permeability, microbiota modifications, and intestinal infection [71]. Furthermore, the upper and lower gastrointestinal tract can show mild to moderate inflammation [72].

More than 90% of ASD children have feeding problems with detrimental effects; one of the causes is the modification of the microbiome, suggesting that ASD behaviors could benefit from interventions to restore microbial balance [73]. Such modification, known as gut dysbiosis, has been investigated by studying the bacterial genus *Clostridium*, which contributes essential species to the human microbiome [74]. Data suggest an association of ASD with an increase in *Clostridium* and a decrease in other microbiome species [75]. Notwithstanding, dysbiosis of other species is also correlated to autism [76]. Beyond the microbiome, several changes occur in enteric neurons and enteric glia [77] that affect the appropriate communication to central structures [78], making the microbiota–gut–brain axis a pivotal center to study the underlying basis of autism.

6.6 Alzheimer's disease (AD)

AD is a neurodegenerative condition showing a progressive deterioration of higher brain functions, mainly memory, and is considered one of the most common dementias [79]. The striking tissue features are the presence of extracellular accumulations of the amyloid beta (Aβ) peptide, or amyloid plaques, and neurofibrillary tangles [80]. Unfortunately, the evidence to explain the correlation between ENS and AD remains scarce. However, AD patients suffer from gastrointestinal alterations [81], that Aβ is also accumulated in enteric neurons producing a number reduction and suggesting that ENS dysfunction is ligated to AD [82] and that gut dysbiosis could also be correlated to AD [83].

6.7 Parkinson's disease (PD)

PD is a neurodegenerative disease with a progressive reduction in the number of dopaminergic neurons in the substantia nigra pars compacta (SN), associated with abnormal cytoplasmic deposits mainly of alpha-synuclein, known as Lewy bodies [84, 85]. It is usually diagnosed by the motor characteristics of the patient, such as progressive tremors, jaw rigidity, and bradykinesia [86]. Data indicate that Lewy bodies are also found in enteric neurons, correlated with gastrointestinal motility and constipation [87].

6.8 Chagasic megacolon

Chagas disease is caused after an infection by *Trypanosoma cruzi*. It is an endemic disease in South America, Central America, and Mexico. The acute symptoms often go unrecognized, but patients can develop many physiological alterations, including motor dysfunction of the gastrointestinal tract [88]. The mechanism of enteric neuronal lesions in the intestinal plexuses generates aperistalsis and megasyndromes. In the case of megacolon, motility problems are associated with colon enlargement and constipation, showing the widening of the luminal region and muscle hypertrophy. Such lesions occur because the infection causes immune reactions that progressively become cytotoxic, producing oxidative stress and a reduction in the number of neurons [89].

7. Conclusions

The ENS is a sophisticated nervous system by itself, with an elaborated organization immersed in the whole gastrointestinal tract, responsible for regulating gut physiology. Its million neurons include intrinsic and extrinsic neural pathways with a significant function independent from the central nervous system. This attribute insinuates that the ENS should be considered as a second brain. To sustain all functions around food processing, the complex network of neurons and glial cells and the relationships with the microbiome organize an exceptional bidirectional communication with the CNS and an intrinsic communication with its own afferents, interneurons, and motor neurons. These pathways are essential for maintaining a healthy gut and general homeostasis. Dysregulation of such networks produces a wide range of diseases. Thus, research on the ENS and its interaction with the CNS must grow to give new insights into the function and pathophysiology of the gastrointestinal tract to develop new and better therapeutic approaches.

Acknowledgements

This chapter was funded by Conacyt Fellowship 1036810 to FMJCS and by funds from the Cuerpos Académicos Neuroscience (UV-CA-28), and Neurochemistry (UV-CA-304) at the Brain Research Institute.

Conflict of interest

The authors declare no conflict of interest.

DOI: http://dx.doi.org/10.5772/intechopen.112571

Author details

Flower M.J. Caycho Salazar[1,2], Deissy Herrera-Covarrubias[2], Genaro A. Coria-Ávila[2], Luis I. García-Hernández[2], María Rebeca Toledo-Cárdenas[2], María Elena Hernández-Aguilar[2] and Jorge Manzo[2]*

1 PhD Program in Brain Research, University of Veracruzana, Xalapa, Veracruz, Mexico

2 Brain Research Institute, University of Veracruzana, Xalapa, Veracruz, Mexico

*Address all correspondence to: jmanzo@uv.mx

References

[1] Steele RE, David CN, Technau U. A genomic view of 500 million years of cnidarian evolution. Trends in Genetics. 2011;**27**(1):7-13

[2] Kinnamon JC, Westfall JA. A three dimensional serial reconstruction of neuronal distributions in the hypostome of a Hydra. Journal of Morphology. 1981;**168**(3):321-329

[3] Furness JB, Stebbing MJ. The first brain: Species comparisons and evolutionary implications for the enteric and central nervous systems. Neurogastroenterology and Motility. 2018;**30**(2):1-6

[4] Neckel PH. Annotated translation of Georg Meissner's first description of the submucosal plexus. Neurogastroenterology and Motility. 2023;**35**(3):e14480

[5] Young HM, Hearn CJ, Newgreen DF. Embryology and development of the enteric nervous system. Gut. 2000;**47**(suppl 4):iv12

[6] Nagy N, Goldstein AM. Enteric nervous system development: A crest cell's journey from neural tube to colon. Seminars in Cell & Developmental Biology. 2017;**66**:94-106

[7] Bronner ME, LeDouarin NM. Development and evolution of the neural crest: An overview. Developmental Biology. 2012;**366**(1):2-9

[8] Memic F, Knoflach V, Morarach K, Sadler R, Laranjeira C, Hjerling-Leffler J, et al. Transcription and signaling regulators in developing neuronal subtypes of mouse and human enteric nervous system. Gastroenterology. 2018;**154**(3):624-636

[9] Popowycz N, Uyttebroek L, Hubens G, Nassauw L van. Differentiation and subtype specification of enteric neurons: Current knowledge of transcription factors, signaling molecules and signaling pathways involved. Journal Cell Signalling. 2022;**1**(3):14-27

[10] Timmermans J, Hens J, Adriaensen D. Outer submucous plexus: An intrinsic nerve network involved in both secretory and motility processes in the intestine of large mammals and humans. The Anatomical Record. 2001;**262**(1):71-78

[11] Hansen MB. The enteric nervous system I: Organisation and Classification. Pharmacology & Toxicology. 2003;**92**(3):105-113

[12] Savulescu-Fiedler I, Gurghean AL, Siliste RN. The complex involvement of the digestive tract in human defense behavior – structural and functional arguments. Journal of Medicine and Life. 2022;**15**(9):1081-1089

[13] Brehmer A. Classification of human enteric neurons. Histochemistry and Cell Biology. 2021;**156**(2):95-108

[14] Stach W. Neuronal organization of the plexus myentericus (Auerbach) in the small intestine of the pig. IV. Type IV-Neurons. Zeitschrift für Mikroskopisch-Anatomische Forschung. 1982;**96**(6):972-994

[15] Stach W. Neuronal organization of the myenteric plexus (Auerbach's) in the pig small intestine. V. Type-V neurons. Zeitschrift für Mikroskopisch-Anatomische Forschung. 1985;**99**(4):562-582

[16] Brehmer A, Schrödl F, Neuhuber W. Correlated morphological and chemical

phenotyping in myenteric type V neurons of porcine ileum. The Journal of Comparative Neurology. 2002;**453**(1):1-9

[17] Furness JB, Bornstein JC, Trussell DC. Shapes of nerve cells in the myenteric plexus of the guinea-pig small intestine revealed by the intracellular injection of dye. Cell and Tissue Research. 1988;**254**(3):561-571

[18] Gabella G. Ultrastructure of the nerve plexuses of the mammalian intestine: The enteric glial cells. Neuroscience. 1981;**6**(3):425-436

[19] Coelho-Aguiar J et al. The enteric glia: Identity and functions. Glia. 2015;**63**(6):921-935

[20] Hanani M, Reichenbach A. Morphology of horseradish peroxidase (HRP)-injected glial cells in the myenteric plexus of the guinea-pig. Cell and Tissue Research. 1994;**278**(1):153-160

[21] Gulbransen BD, Sharkey KA. Novel functional roles for enteric glia in the gastrointestinal tract. Natural Review in Gastroenterology. 2012;**9**(11):625-632

[22] Baquero F, Nombela C. The microbiome as a human organ. Clinical Microbiology and Infection. 2012;**18**(s4):2-4

[23] Eckburg PB, Bik EM, Bernstein CN, Purdom E, Dethlefsen L, Sargent M, et al. Diversity of the human intestinal microbial flora. Science. 2005;**308**(5728):1635-1638

[24] Palmer C, Bik EM, DiGiulio DB, Relman DA, Brown PO. Development of the human infant intestinal microbiota. PLoS Biology. 2007;**5**(7):e177

[25] Barko PC, McMichael MA, Swanson KS, Williams DA. The Gastrointestinal microbiome: A review. Journal of Veterinary Internal Medicine. 2018;**32**(1):9-25

[26] Amon P, Sanderson I. What is the microbiome? Archives of Disease in Childhood. Education and Practice Edition. 2017;**102**(5):257

[27] Dockray GJ. Luminal sensing in the gut: An overview. Journal of Physiology Pharmacology Official Journal of Polish Physiological Society. 2003;**54** (Suppl. 4):9-17

[28] Dupont HL, Jiang ZD, Dupont AW, Utay NS. The intestinal microbiome in human health and disease. Transactions of American Clinical Climate Association. 2020;**131**:178-197

[29] Bertrand PP, Thomas EA. Multiple levels of sensory integration in the intrinsic sensory neurons of the enteric nervous system. Clinical Experiment in Pharmacology. 2004;**31**(11):745-755

[30] Giuffrè M, Moretti R, Campisciano G, Silveira ABM, et al. You talking to me? Says the enteric nervous system (ENS) to the microbe. How intestinal microbes interact with the ENS. Journal of Clinical Medicine. 2020;**9**(11):3705

[31] Furness JB, Jones C, Nurgali K, Clerc N. Intrinsic primary afferent neurons and nerve circuits within the intestine. Progress in Neurobiology. 2004;**72**(2):143-164

[32] Hansen MB. The enteric nervous system II: Gastrointestinal functions: The enteric nervous system and gastrointestinal functions. Pharmacology & Toxicology. 2003;**92**(6):249-257

[33] Atsumi K, Yajima T, Tachiya D, Kokubun S, Shoji N, Sasano T, et al. Sensory neurons in the human jugular ganglion. Tissue & Cell. 2020;**64**:101344

[34] Blackshaw LA, Brookes SJH, Grundy D, Schemann M. Sensory transmission in the gastrointestinal tract. Neurogastroenterology and Motility. 2007;**19**(s1):1-19

[35] Keller JT, Beduk A, Saunders MC. Central brainstem projections of the superior vagal ganglion of the cat. Neuroscience Letters. 1987;**75**(3):265-270

[36] Browning KN, Mendelowitz D. Musings on the wanderer: What's new in our understanding of vago-vagal reflexes?: II. Integration of afferent signaling from the viscera by the nodose ganglia. American Journal of Physiology. Gastrointestinal and Liver Physiology. 2002;**284**(1):G8-G14

[37] Powley TL. Brain-gut communication: Vagovagal reflexes interconnect the two "brains". American Journal of Physiol-Gastric L. 2021;**321**(5):G576-G587

[38] Gasparini S, Howland JM, Thatcher AJ, Geerling JC. Central afferents to the nucleus of the solitary tract in rats and mice. The Journal of Comparative Neurology. 2020;**528**(16):2708-2728

[39] Grundy D. Neuroanatomy of visceral nociception: Vagal and splanchnic afferent. Gut. 2002;**51**(suppl 1):i2

[40] Berthoud HR, Blackshaw LA, Brookes SJH, Grundy D. Neuroanatomy of extrinsic afferents supplying the gastrointestinal tract. Neurogastroenterology and Motility. 2004;**16**(s1):28-33

[41] Bharucha AE. Pelvic floor: Anatomy and function. Neurogastroenterology and Motility. 2006;**18**(7):507-519

[42] Brierley SM, Hibberd TJ, Spencer NJ. Spinal afferent innervation of the colon and rectum. Frontiers in Cellular Neuroscience. 2018;**12**:467

[43] Costa M, Furness JB. The peristaltic reflex: An analysis of the nerve pathways and their pharmacology. Naunyn-Schmiedeberg's Archives of Pharmacology. 1976;**294**(1):47-60

[44] Bornstein JC, Costa M, Grider JR. Enteric motor and interneuronal circuits controlling motility. Neurogastroenterology and Motility. 2004;**16**(s1):34-38

[45] Lomax AE, Linden DR, Mawe GM, Sharkey KA. Effects of gastrointestinal inflammation on enteroendocrine cells and enteric neural reflex circuits. Autonomic Neuroscience. 2006;**126**:250-257

[46] Gershon MD. Nerves, reflexes, and the enteric nervous system. Journal of Clinical Gastroenterology. 2005;**39**(5):S184-S193

[47] Travagli RA, Anselmi L. Vagal neurocircuitry and its influence on gastric motility. Natural Review in Gastroenterology. 2016;**13**(7):389-401

[48] Browning KN, Travagli RA. Plasticity of vagal brainstem circuits in the control of gastric function. Neurogastroenterology and Motility. 2010;**22**(11):1154-1163

[49] Obata Y, Pachnis V. The effect of microbiota and the immune system on the development and organization of the enteric nervous system. Gastroenterology. 2016;**151**(5):836-844

[50] Muller PA, Koscsó B, Rajani GM, Stevanovic K, Berres ML, Hashimoto D, et al. Crosstalk between muscularis macrophages and enteric neurons regulates gastrointestinal motility. Cell. 2014;**158**(5):1210

[51] Okabe Y, Medzhitov R. Tissue-specific signals control reversible program of localization and functional polarization of macrophages. Cell. 2014;**157**(4):832-844

[52] Phillips RJ, Powley TL. Macrophages associated with the intrinsic and extrinsic autonomic innervation of the rat gastrointestinal tract. Autonomic Neuroscience. 2012;**169**(1):12-27

[53] Gabanyi I, Muller PA, Feighery L, Oliveira TY, Costa-Pinto FA, Mucida D. Neuro-immune interactions drive tissue programming in intestinal macrophages. Cell. 2016;**164**(3):378-391

[54] Meroni E, Stakenborg N, Viola MF, Boeckxstaens GE. Intestinal macrophages and their interaction with the enteric nervous system in health and inflammatory bowel disease. Acta Physiologica (Oxford, England). 2019;**225**(3):e13163

[55] Holland AM, Bon-Frauches AC, Keszthelyi D, Melotte V, Boesmans W. The enteric nervous system in gastrointestinal disease etiology. Cellular and Molecular Life Sciences. 2021;**78**(10):4713-4733

[56] Gazouli M, Wouters MM, Kapur-Pojskić L, Bengtson MB, Friedman E, Nikčević G, et al. Lessons learned – resolving the enigma of genetic factors in IBS. Natural Review in Gastroenterology. 2016;**13**(2):77-87

[57] Lacy BE, Patel NK. Rome criteria and a diagnostic approach to irritable bowel syndrome. Journal of Clinical Medicine. 2017;**6**(11):99

[58] Niesler B, Kuerten S, Demir IE, Schäfer KH. Disorders of the enteric nervous system – a holistic view. Nature Review in Gastroenterology. 2021;**18**(6):393-410

[59] Corsetti M, Oudenhove LV, Tack J. The quest for biomarkers in IBS – where should it lead us? Neurogastroenterology and Motility. 2014;**26**(12):1669-1676

[60] Spiller R. Irritable bowel syndrome: New insights into symptom mechanisms and advances in treatment. F1000research. 2016;**5**:780

[61] Ng QX, Soh AYS, Loke W, Lim DY, Yeo WS. The role of inflammation in irritable bowel syndrome (IBS). Journal of Inflammation Research. 2018;**11**:345-349

[62] Tjaden NEB, Trainor PA. The developmental etiology and pathogenesis of Hirschsprung disease. Translational Research. 2013;**162**(1):1-15

[63] Martucciello G. Hirschsprung's disease as a neurochristopathy. Pediatric Surgery International. 1997;**12**(1):2-10

[64] Boeckxstaens GE, Zaninotto G, Richter JE. Achalasia. Lancet. 2014;**383**(9911):83-93

[65] Herbella FAM, Aquino JLB, Stefani-Nakano S, Artifon ELA, Sakai P, Crema E, et al. Treatment of achalasia: Lessons learned with Chagas' disease. Diseases of the Esophagus. 2008;**21**(5):461-467

[66] Zilberstein B, Cleva R, Gabriel AG, Neto SG, Gama-Rodrigues JJ. Congenital achalasia: Facts and fantasies. Diseases of the Esophagus. 2005;**18**(5):335-337

[67] Barberio B, Judge C, Savarino EV, Ford AC. Global prevalence of functional constipation according to the Rome criteria: A systematic review and meta-analysis. The Lancet Gastroenterology & Hepatology. 2021;**6**(8):638-648

[68] Sharma A, Rao S. Constipation: Pathophysiology and current therapeutic

approaches. Handbook of Experimental Pharmacology. 2017;**239**:59-74

[69] Wang J, Ma B, Wang J, Zhang Z, Chen O. Global prevalence of autism spectrum disorder and its gastrointestinal symptoms: A systematic review and meta-analysis. Frontiers in Psychiatry. 2022;**13**:963102

[70] Manzo J. Un segundo espectro del autismo: de la conducta a la neurona. eNeurobiol. 2019;**23**(10):1501

[71] Ristori MV, Quagliariello A, Reddel S, Ianiro G, Vicari S, Gasbarrini A, et al. Autism, gastrointestinal symptoms and modulation of gut microbiota by nutritional interventions. Nutrients. 2019;**11**(11):2812

[72] Horvath K, Perman JA. Autism and gastrointestinal symptoms. Current Gastroenterology Reports. 2002;**4**(3):251-258

[73] Mulle JG, Sharp WG, Cubells JF. The gut microbiome: A new frontier in autism research. Current Psychiatry Reports. 2013;**15**(2):337

[74] Marathe NP, Shetty SA, Lanjekar VB, Rasane MH, Ranade DR, Shouche YS. Genome sequencing of multidrug resistant novel Clostridium sp. BL8 reveals its potential for pathogenicity. Gut Pathogens. 2014;**6**(1):30

[75] Argou-Cardozo I, Zeidán-Chuliá F. Clostridium bacteria and autism spectrum conditions: A systematic review and hypothetical contribution of environmental glyphosate levels. Medical Science. 2018;**6**(2):29

[76] Rosenfeld CS. Microbiome disturbances and autism spectrum disorders. Drug Metabolism and Disposition. 2015;**43**(10):1557-1571

[77] Grubišić V, Parpura V. The second brain in autism spectrum disorder: Could connexin 43 expressed in enteric glial cells play a role? Frontiers in Cellular Neuroscience. 2015;**9**:242

[78] Theije CGM de, Wu J, Silva SL da, Kamphuis PJ, Garssen J, Korte SM, et al. Pathways underlying the gut-to-brain connection in autism spectrum disorders as future targets for disease management. European Journal of Pharmacology. 2011;**668**:S70-S80

[79] Castillo-Rangel C, Marín G, Diaz-Chiguer DL, Zarate-Calderon CJ, Viveros-Martinez I, Caycho-Salazar FDMDJ, et al. Animal models in Alzheimer's disease: Biological plausibility and mood disorders. Neurology Perspectives. 2023;**3**(1):100110

[80] Lane CA, Hardy J, Schott JM. Alzheimer's disease. European Journal of Neurology. 2018;**25**(1):59-70

[81] Chalazonitis A, Rao M. Enteric nervous system manifestations of neurodegenerative disease. Brain Research. 2018;**1693**(Pt B):207-213

[82] Rao M, Gershon MD. The bowel and beyond: The enteric nervous system in neurological disorders. Nature Review in Gastroenterology. 2016;**13**(9):517-528

[83] Barrio C, Arias-Sánchez S, Martín-Monzón I. The gut microbiota-brain axis, psychobiotics and its influence on brain and behaviour: A systematic review. Psychoneuroendocrinology. 2022;**137**:105640

[84] Shults CW. Lewy bodies. Proceedings of the National Academy of Science. 2006;**103**(6):1661-1668

[85] Simon DK, Tanner CM, Brundin P. Parkinson Disease epidemiology, pathology, genetics, and pathophysiology. Clinics in Geriatric Medicine. 2020;**36**(1):1-12

[86] Stoessl AJ, Rivest J. Differential diagnosis of parkinsonism. The Canadian Journal of Neurological Sciences. 1999;**26**(S2):S1-S4

[87] Ohlsson B, Englund E. Atrophic myenteric and submucosal neurons are observed in Parkinson's disease. Parkinsons Disease. 2019;**2019**:7935820

[88] Goldstein AM, Thapar N, Karunaratne TB, Giorgio RD. Clinical aspects of neurointestinal disease: Pathophysiology, diagnosis, and treatment. Developmental Biology. 2016;**417**(2):217-228

[89] Silveira ABM da, Lemos EM, Adad SJ, Correa-Oliveira R, Furness JB, Reis DD. Megacolon in Chagas disease: A study of inflammatory cells, enteric nerves, and glial cells. Human Pathology. 2007;**38**(8):1256-1264

Section 4

Neurological Insights into Cardiovascular and Neuromuscular Conditions

Chapter 7

Diabetic Cardiac Autonomic Neuropathy: Link between Heart Rate Variability, Violated Blood Pressure Pattern, and Pulse Wave Velocity

Victoria Serhiyenko, Marta Hotsko, Yuriy Markevich, Martyn-Yurii Markevich, Volodymyr Segin, Ludmila Serhiyenko and Alexandr Serhiyenko

Abstract

Abnormalities in heart rate (HR) variability (HRV) and blood pressure (BP) variability may increase the risk of cardiovascular diseases. A well-known risk factor for cardiovascular morbidity, such as arrhythmias, stroke, congestive heart failure, heart attacks, and sudden death syndrome, is cardiac autonomic neuropathy (CAN). It has been claimed that chronobiologically evaluating HRV and BP and optimizing timed treatment efficacy can significantly lower the risk of cardiac or stroke death. Physiological cardiovascular activities are under the control of the cardiac autonomic nervous system. Damage of the autonomic nerves leads to dysfunction in HR control and vascular dynamics, notably to CAN. For people with diabetes mellitus (DM), metabolic abnormalities and significant morbidity and mortality are caused by an autonomic imbalance between the sympathetic and parasympathetic nervous systems, which regulate cardiovascular function. There is a strong correlation between changes in neuroendocrine sleep architecture, circadian clock oscillations, glucose metabolism, autonomic function, and diurnal profiles of BP and HR, and there has been evidence of circadian rhythm misalignment in DM patients. The purpose of the chapter is to analyze the current state of the problem in the relationship between DM and circadian rhythm disorders, HRV, and arterial stiffness.

Keywords: diabetes mellitus, autonomic nervous system, diabetic cardiac autonomic neuropathy, heart rate variability, arterial stiffness monitoring

1. Introduction

Many biological activities rely on the circadian clock as the main regulator of metabolism and energy homeostasis. Dyslipidemia (DLP), insulin resistance (IR), and hyperglycemia are all seen in animal models, where the suprachiasmatic nucleus (SCN) of the hypothalamus has been altered. Circadian disorders, such as a decline in the sleep-wake cycle brought on by insufficient sleep, shift work, and social jet lag, have been linked to symptoms of the metabolic syndrome (MeTs) such as impaired glucose tolerance (IGT), insulin sensitivity, hypertriglyceridemia, an increase in body mass index (BMI), and mean arterial blood pressure (BP) [1, 2]. Taken together, these studies suggest that autonomic nervous system (ANS) dysfunction could play a role in the pathogenesis of glucose dysregulation [3]. According to physiological principles, insulin is directly secreted from β-cells under the control of the parasympathetic nervous system (PNS), and the body's glucose levels are managed by the ANS [4]. Although autonomic dysfunction is linked to an increased risk of cardiovascular diseases (CVDs), the specific mechanism by which autonomic dysfunction is linked to CVDs is uncertain [5, 6]. Reduced heart rate variability (HRV) and reduced baroreflex sensitivity (BRS) are early indicators of cardiac autonomic dysfunction [6–9], cardiac attacks, congestive cardiac failure, stroke, and sudden arrhythmic death are all very susceptible to cardiac autonomic neuropathy (CAN).

In the latter, hypertension (HTN) is followed by structural remodeling of the myocardium, including fibrosis and hypertrophy.

This remodeling is accompanied by changes in the extracellular matrix composition, as well as changes in the expression, distribution, and function of cell membrane ion channels, Ca2-cycling proteins, and intercellular gap junction connexin-43 channels [8–10].

Several epidemiological studies have found that greater arterial stiffness, independent of other cardiovascular risk factors, predicts mortality and morbidity. Diabetes can aggravate arterial stiffness by causing pathological changes in the vascular bed, such as changes in the type or structure of collagen and/or elastin in the arterial wall, decreased nitric oxide (NO) bioavailability, chronic low-grade inflammation, increased oxidative stress (OS), and increased sympathetic tone [11].

Hyperglycemia, which is linked to the activation of pro-inflammatory transcription factors and an increase in OS, which results in vasculopathy, is responsible for many of the pathophysiological mechanisms underlying vascular dysfunction in diabetes mellitus (DM). Elevated levels of advanced glycation end products (AGEs) may affect the molecular matrix of the vessel wall.

According to certain research, type 2 diabetes (T2DM) may reduce endothelial NO bioavailability and attenuate vascular smooth muscle cells (VSMCs) response to NO by causing endothelium and VSMC to malfunction in diabetics compared to controls. All of these mechanisms are implicated in mediating hyperglycemia-induced arterial stiffness.

When compared to healthy, age- and sex-matched controls, patients with primary autonomic failure who do not have DM had stiffer aortas.

According to these findings, there is a pathophysiological link between arterial stiffness and cardiac autonomic dysfunction, and ANS is critical for preserving the elastic characteristics of the arteries [12].

Advanced glycation end products (AGEs) production, protein kinase C activation, low-grade inflammation, and endothelial dysfunction are all shared pathogenetic pathways that link arterial stiffness with cardiac autonomic dysfunction [13, 14]. It is still unknown whether increased arterial stiffness causes impaired cardiac autonomic function or whether impaired cardiac autonomic function causes arterial stiffness, as well as the pathophysiological relationship between arterial stiffness and autonomic dysfunction.

2. Type 2 diabetes mellitus

Diabetes and changes in cardiac autonomic control have a well-established link. This includes individuals with IR and children of T2DM patients who have never had diabetes or HTN, where the mean values of HRV are reduced. In healthy individuals, low HRV is linked to a rise in the occurrence of this illness. Numerous research findings suggest a tight connection between T2DM and autonomic neuropathy. Increasing research suggests that CAN is widespread in people with IGT. A metabolic insult and obesity may trigger the onset of CAN. In addition, individuals with IGT frequently experience autonomic symptoms. In addition to CAN, people with IGT have abnormalities in endothelium peripheral vasoreactivity and sudomotor function [15].

Liu et al. retrospectively reviewed 104 patients with T2DM and coronary heart disease (CHD) medical records. Correlation analyses were carried out between HRV measures, clinical parameters, and the severity of coronary lesions. The Gensini scores and the number of damaged arteries were used to determine the severity of coronary lesions. According to Spearman's correlation analysis, there is a substantial negative association between the standard deviation of 5-min mean intervals of NN (SDANN) scores, a component of HRV, and Gensini scores, which remained significant after adjustment for clinical covariates. The research has demonstrated that in patients with T2DM the overall connection between CAN and coronary lesions may exist independently of established variables in the etiology of vascular endothelial damage and atherosclerosis [16, 17]. These findings suggest that endothelial dysfunction and cardiac autonomic nervous dysfunction are related pathophysiologically in T2DM [18]. Reduced HRV values are related to the severity of coronary artery lesions among persons with stable angina pectoris [19]. Liu et al. indicated that CAN might reflect the progression of coronary atherosclerosis in persons with T2DM. According to the authors, CAN may be associated with the degree of coronary atheromatous burden in T2DM patients. The combination of cardiac autonomic nervous function testing and plaque enhancement may improve CHD risk classification in T2DM patients [16].

CAN results from impaired autonomic function and subsequent nervous system imbalance of the cardiovascular system that occurs due to diabetes [20]. CAN can also result from metabolic disturbances in prediabetes and MeTs, which are conditions before established diabetes [8]. If a person develops type 1 diabetes (T1DM), hyperglycemia affects several cellular pathways, leading to microvascular problems, including Cardiac autonomic neuropathy (CAN) [6]. A complicated link between rising IR and developing autonomic dysfunction leads to CAN in prediabetes, MeTs, and T2DM [21]. Diabetes patients who have high blood glucose have the following changes: positive regulation of the hexosamine pathway, which leads to an increase in N-acetyl glucosamine and, consequently, the induction of OS; exacerbation of OS due to lipid peroxidation and reduction of glutathione levels and enzymes involved in antioxidant defense; increase in sorbitol, whose intracellular increase promotes osmotic stress and greater electrolyte output from the cell, which causes impairment of Schwann cells from peripheral neurons [22]. The primary driver for CAN development in T1DM is hyperglycemia compared with the multifactorial pathogenesis of CAN in T2DM. In early CAN, IR directly promotes sympathetic predominance [21]. T2DM patients have a variety of vascular risk factors, including DLP and HTN, which can lead to microvascular disease [9]. Diabetes patients with low HRV have an increased risk of complications and mortality compared to those with normal HRV values. SampEn (a nonlinear measure used to evaluate the regularity of a time series) and high-frequency (HF) power of HRV, among other metrics, are superior discriminators for detecting autonomic dysfunction [23]. People with diabetes had a lower amplitude of day-night fluctuations in HRV from a circadian standpoint. These data support using HRV as a risk indicator in T2DM [10, 24].

3. The circadian rhythms and heart rate variability

Circadian rhythms are controlled by circadian clocks, which control day-night oscillations for ~24 hours. The molecular mechanism of the circadian clock is based on a negative feedback loop, which includes four main transcription factors of the helix-loop-helix or per-arnt-sim domain: CLOCK, BMAL1, PER, and CRY3 [25]. The SCN clock controls the circadian rhythm of the heart rate (HR) through circadian changes in the autonomic tone of the sinus node, particularly increased vagal tone at night.

The SCN regulates the diurnal release of other neurohumoral substances, but its effect on the HR's circadian rhythm is unknown. Two mechanisms are responsible for the circadian rhythm of HR: the central circadian clock in the SCN of the hypothalamus can directly influence the electrophysiology of the heart and arrhythmogenesis through various factors, including the ANS and a local circadian clock within the heart (albeit controlled by a central clock) can regulate the circadian rhythm of ion channel expression in the heart. The ANS can play a unique role as a bridge between clocks. In particular, the ANS is capable of synchronizing the local clock in order to effectively control fluctuations in the expression of ion channels [25].

The ANS reacts to internal and external stimuli and maintains the organism's homeostasis. The autonomic background of the cardiac periodicity control is important for the interpretation of HRV measurements. Under physiologically normal conditions, the respiratory rate is in the HF range, which relates the HF components to the vagal modulations of the heart period. At the same time, the low-frequency (LF) power of HRV modulations reflects a combined vagal and sympathetic control [26]. Because the sympathetic and vagal systems induce different frequencies of cardiac periodicity modulation, the HRV is an appropriate approach for analyzing the relative strength of modulations by both limbs of the autonomic system. The latter makes it possible to assess the relationship between sympathetic and vagal control using the LF/HF ratio or the so-called normalized LF and HF components (excluding very low-frequency components) [26]. Because of the relationship between cardiac ANS status and spectrum components of HRV (and other analogs of measurement), HRV analysis is useful for diagnosing autonomy, which is defined by a lack of autonomic response. Numerous studies have demonstrated the clinical utility of HRV in the early diagnosis and type classification of diabetic neuropathy [26, 27].

The clinical significance of HRV is determined by the following provisions [28]:

- First, HRV can detect early subclinical manifestations of autonomic dysfunction, which could be useful from a therapeutic standpoint in understanding the subject's risk and subsequent care. In other words, having knowledge into HRV could influence the aggressiveness of the treatment and the choice of treatment when dealing with hyperglycemia and complications, as well as recognizing possible hazards that are not visible (*e.g.*, CAN).

- Second, T2DM management must be comprehensive and include preventive intervention.
- Third, a better understanding of interventions that could improve HRV may allow guiding the patient toward lifestyle changes that can improve HRV parameters and, therefore, quality of life.

4. Autonomic nervous system and heart rate variability

It is common knowledge that the ANS is a component of the peripheral nervous system that regulates involuntary physiologic processes, including HR, BP, respiration, digestion, and sexual arousal [29]. The ANS regulates cardiac function by balancing the effects of sympathetic and parasympathetic nervous system impulses on the heart. These autonomic signals are integrated by intrinsic cardiac neurocircuits to fine-tune cardiac regulation, and sensory feedback loops regulate autonomic transmission in response to external stimuli [13]. The sympathetic nervous system (SNS) is responsible for activating activity and attention: the "fight or flight" response. During this process, BP and HR rise, glycogenolysis occurs, etc. Almost every living tissue in the body is innervated by the SNS. The PNS encourages "rest and digest" processes, HR and BP drop, gastrointestinal peristalsis, etc. The ANS swiftly governs and modifies BP and HR interaction through the arterial baroreflex. A negative feedback loop buffers change in HR and BP to maintain them under various situations in daily life [29, 30]. The ANS dynamically controls the heart. Control of chronotropy, lusitropy, dromotropy, and inotropy is handled by a hierarchical neural network. Intrinsic autonomic dysfunction is caused by disorders that directly impact the autonomic nerves, such as DM and primary autonomic failure syndromes. Extrinsic autonomic dysfunction reflects alterations in autonomic function caused by cardiac or other illness. Therefore, these two types of autonomic dysfunction can be associated with diabetes [31].

Strong physiological connections link the ANS with glucose metabolism. A network of autonomic nerve fibers surrounds pancreatic islet cells and their blood vessel supply. In healthy individuals, parasympathetic nerve signaling triggers the early release of insulin from the pancreatic β-cells (*e.g.*, first-phase insulin release) in response to sensory signals [32].

In rodent models, PNS increases β-cell proliferation [33]. In response to hypoglycemia, sympathetic nerve transmission modulates glucagon secretion by islet α-cells and suppresses insulin secretion [4, 9]. β-adrenoceptors expressed in α-cells and α2-adrenoceptors expressed in β-cells are responsible for these two SNS actions. The ANS also innervates the liver, adipose tissue, and smooth muscle tissue, whose effects on glucose metabolism are related to insulin sensitivity [3, 34].

Autonomic dysfunction, as measured by decreased HRV, has also been linked to calcification of the coronary arteries [35, 36]. Whether dysfunction of the ANS contributes to the development of atherosclerosis, it will significantly impact our understanding of the pathogenesis of coronary atherosclerosis in patients with diabetes. A direct effect of autonomic disorders on atherosclerosis is plausible. Sympathetic denervation may result in VSMCs dedifferentiation and a shift to a phenotype associated with extracellular matrix synthesis and migration to the intima, and changes seen in atherosclerosis [37].

5. Heart rate variability and arterial stiffness

In T2DM, cardiac autonomic dysfunction induces HR increase, resulting in diastole shortening; however, it appears that independent of the effect on HR, cardiac autonomic dysfunction can shorten diastole duration (DD). Because DD has a significant impact on subendocardial myocardial viability (SVI) and cardiac autonomic dysfunction, in addition to arterial stiffness, plays a main role in SVI impairment and may thus affect cardiovascular prognosis.

In T2DMs, cardiac autonomic dysfunction causes HR to accelerate, resulting in diastole shortening; however, it appears that cardiac autonomic dysfunction may decrease DD regardless of the effect on HR. Since DD strongly influences SVI, cardiac autonomic dysfunction plays a primary role in addition to arterial stiffness in the impairment of SVI and as a result, the cardiovascular prognosis may worsen [38, 39].

Reduced HRV in uncomplicated diabetics indicates an obscure process of autonomic neuropathy in diabetic patients, which occurs even before clinical atherosclerotic CVD is manifested. Surrogate atherosclerosis indicators have also been linked to lower HRV, and higher carotid intima-media thickness (CIMT) in T2DM patients has been linked to lower HRV, independent of conventional cardiovascular risk factors. As a result, the existence of CAN should be considered quite early in the course of diabetes rather than after clinical CVD develops [40, 41].

Chorepsima et al. discovered that, in addition to BP, decreased cardiac autonomic function as measured by HRV was a significant predictor of abnormal pulse wave velocity (PWV) in T2DM patients. Moreover, lower HRV values were independently related to higher PWV [12]. A decrease in HRV has been associated to an increased risk of death in people with CHD or diabetes. HRV can also be used to detect BRS control, specifically vagal control. As a result, vascular stiffness may influence BRS and, as a result, HRV. Increased arterial stiffness, as assessed by PWV and/or the ambulatory arterial stiffness index, has been associated to coronary atherosclerosis and a worse cardiovascular prognosis in both the general population and specific disease groups, most notably DM.

Reduced HRV in uncomplicated patients with DM reveals the enigmatic process of CAN in diabetic patients, which begins prior to clinical atherosclerotic CVDs manifest. Surrogate atherosclerosis indicators have also been linked to reduced HRV and greater CIMT has been associated to reduced HRV in T2DM patients, independent of conventional cardiovascular risk factors. As a result, rather than waiting until clinical CVDs have developed, the occurrence of CAN should be evaluated much earlier in the course of DM [40, 42].

Based on Bagherzadeh et al. findings and previous works, one could suggest that atherosclerosis, both due to diabetes and increased age, is influenced by the CAN, resulting in an increased risk of CVD and related mortality in T2DM patients. Compared with normal controls, Bagherzadeh et al. observed increased arterial stiffness and decreased HRV in uncomplicated T2DM patients. The association between HRV indices and PWV was significant after diabetes adjustment; however, this impact was abolished following adjustment for confounders. Based on the findings of this study, it appears that there is a connection between HRV and arterial stiffness as a measure of atherosclerosis in diabetic patients, albeit the influence of confounding factors should be considered [40].

Technically, baroreceptor/reflex sensitivity parameters quantify the HRV change concerning BP variability. BRS represents the ability of carotid sinus and aortic arch baroreceptors to detect changes in aortic distention and communicate them to brainstem nuclei responsible for cardiac autonomic regulation [43]. Parasympathetic discharge is controlled by the nucleus of tractus solitary, while sympathetic outflow is controlled by the rostral ventrolateral medulla [44]. In actuality, the intensity of the link between changes in BP and reflex modulation of HR can be determined by two checkpoints mediated by the status of baroreceptors. The first corresponds to the physical features of the vascular system, which transports pressure signals to baroreceptors, while the second represents the ANS's primary effector. The latter is known as neuronal BRS, whereas the former is known as mechanical BRS. Autonomic neuropathy, rather than carotid stiffness, has been found to be a greater predictor of BRS in T2DM patients [44]. To this end, different studies have focused on characterizing changes in the different ganglia and nuclei in the cardiac neural network in MeTs, prediabetes, and T2DM [10].

Arterial stiffness is a marker of subclinical vascular disease that has been documented among young patients with T1DM. The exposure of blood vessels to the continuous harmful effects of hyperglycemia may result in the early development of arterial stiffness in these patients. Medial arterial calcification (MAC), a complicated dynamic process influenced by multiple molecular signaling pathways, causes peripheral arterial stiffness. Despite peripheral arterial stiffness is known to be linked to peripheral artery disease (PAD) [45], the impact of MAC on peripheral vascular beds is poorly understood, despite the fact that advanced stages of MAC, which are characterized by loss of elasticity in arterial walls, are associated with worse tissue perfusion, ultimately leading to arterial flow stasis. MAC was formerly thought to be a harmless illness, despite the presence of associated PAD in a large fraction of patients, at least patients with T1DM. Arterial wall stiffness is now recognized as a key risk factor for cardiovascular death and morbidity, as well as a strong, independent predictor of all-cause mortality, and future CV events [45, 46].

Arterial stiffness is a significant risk factor for cardiovascular disease linked to isolated systolic HTN and elevated pulse pressure (PP) in target organ microvasculature [11]. Mechanotransduction, mitochondrial OS, DLP, a decreased elastin/collagen ratio, production of elastin cross-linking, reactive oxygen species (ROS)-induced inflammation, calcification, VSMC stiffness, endothelial dysfunction, genome mutations, and epigenetics all contribute to artery wall stiffening. Targeting these several biochemical pathways at various times of CVD risk factor exposure could be a novel approach to developing medicines that lower arterial stiffness without impacting artery strength or normal remodeling [47].

Lacolley et al. identify the following key provisions related to arterial stiffness:

- The connection between hemodynamics and mechanosensing is indicated by arterial stiffness.

- In recent years, the processes behind arterial stiffness have changed from elastin and collagen to VSMC phenotypic changes linked to metabolic, genetic, and epigenetic parameters, OS, and mechanotransduction.

- Different cardiovascular (CV) risk factors, such as aging, HTN, DM, and chronic kidney disease (CKD), as well as their various durations of exposure, share the processes that increase arterial stiffness to various levels.

- To combat or perhaps reverse this complex process, it is currently difficult to find medications that target either the early or late stages of arterial stiffness [47].

Numerous pathophysiological research have shown how arterial stiffness is affected by CV risk factors. The processes by which these CV risk factors harden the major arteries have been gradually uncovered [47, 48]. In particular, for people with essential HTN, significant arterial stiffness is raised in response to the bio-mechanical fatigue of the stiff wall materials, such as collagen, brought on by repetitive pulsatile stress and the increased loading of these materials by high blood pressure [49]. Furthermore, the activation of the renin-angiotensin-aldosterone pathway contributes to structural changes in the artery wall *via* VSMC proliferation, low-grade inflammation, increased collagen content, and AGE production. Another example is T2DM, which might harm the large artery wall due to its primary features, which include hyperglycemia and IR. Both variables may exert structural and functional effects *via* various methods. Chronic hyperglycemia stimulates VSMC proliferation and increases the generation of AGEs and collagen cross-linking, stiffening the arterial wall material. Moreover, the expression of matrix metalloproteinase-2 and -9, as well as angiotensin II receptors, is elevated in vascular tissue [47]. Insulin resistance stimulates collagen production and raises the expression of numerous inflammatory-related genes. Arterial stiffness is, thus, most likely a result of these alterations. The final example is CKD, which causes calcifications the large arteries wall. The sequence of molecular processes causing vascular calcification may start with the loss of expression by VSMCs of constitutive inhibitory proteins and end with expression by VSMCs and macrophages of osteoblastic, chondrocytic, and osteoclastic-associated proteins that orchestrate the calcification process [47, 50].

In a community-based cohort study, Zheng et al. discovered that arterial stiffness could be a risk factor for diabetes independent of established risk factors (*e.g.*, age, BMI, BP, and alcohol use). The temporal analysis results, in particular, revealed that a change in arterial stiffness might precede a change in fasting blood glucose (FBG) rather than *vice versa* [49]. Increased arterial stiffness is a common sign of atherosclerotic vascular system involvement and is known to occur as a result of atherosclerotic risk factors such as aging, DLP, HTN, DM, and smoking. CHD, cerebrovascular disease, and PAD are all linked to increased arterial stiffness [38]. These findings indicate that diabetes candidates have endothelial dysfunction and inflammation as their blood glucose levels rise after being diagnosed with T2DM. Endothelial dysfunction is a well-known contributor to the macrovascular and microvascular consequences of diabetes. Endothelial dysfunction may precede diabetes by promoting IR and glucose dysregulation, eventually leading to diabetes. The precise importance of endothelial dysfunction for people with prediabetes and diabetes, however, is still being discussed [38].

According to several studies, increased big arterial stiffness appears as early as prediabetes. Carotid-femoral PWV (cfPWV) was higher in persons with impaired FBG or IGT compared to those with normal glucose metabolism in the ADDITION-Leicester cohort. The rise was identical to that seen in people with newly diagnosed T2DM. Carotid stiffness scores increased with FBG, insulin, and glycated hemoglobin A1c (HbA1c); cfPWV increased with HbA1c, FBG, homeostasis model assessment of IR (HOMA-IR) index, and HbA1c, as well as waist circumference, triglycerides (TG), and high-density lipoprotein cholesterol (HDL-C), were all predictors of cfPWV following a 17-year follow-up. Finally, a recent study in middle-aged CVD-free patients found a link between cfPWV and telomere length (inverse) and the HOMA-IR index (direct), showing that IR associated with chronic inflammation may promote telomere shortening and vascular aging [51].

Cardiovascular autonomic dysfunction has been found in patients with T1DM and lower-extremity arterial calcification or markers of arterial stiffness. The Pittsburgh Epidemiology of Diabetes Complications study found that CAN, as defined by abnormal HRV after deep breathing, was linked with arterial stiffness regardless of traditional CV risk markers in a cohort of 144 individuals with childhood-onset T1DM. In addition, the SEARCH cardiovascular disease project used a cross-sectional design to investigate the correlations between HRV and multiple measures of arterial stiffness in young individuals with T1DM and healthy controls. CAN was linked to arterial stiffness in both the central and peripheral vascular beds, irrespective of other traditional CV risk factors as obesity-related parameters, blood pressure, lipid profiles, smoking, and microalbuminuria.

In addition, aberrant HRV during timed deep breathing was recently linked to aortic stiffness in a large sample of T1DM patients from the Steno Diabetes Center, even after controlling for gender, age, BP, glycemic management, diabetes duration, and renal function [45].

Some studies looked at arterial stiffness measurements in aged individuals and T2DM patients to support the use of these measurements as markers for primary prevention in target populations. A case-control research was carried out to examine PWV and the augmentation index (AIx) in two groups of cardiovascular patients: T2DM and CHD. After adjusting for gender and age, BP and HR discovered a strong link between CHD, PWV, and Aix. Interestingly, when the results of patients with T2DM were compared to healthy individuals, the elevated PWV values remained significant, implying that the outcomes depended on the methods used to quantify arterial stiffness.

A recent outstanding review detailed the potential clinical implications of arterial stiffness on the microvasculature. In summary, arterial stiffness may play a role in the development of numerous brain (*e.g.*, dementia and cognitive impairment), heart (*e.g.*, ischemia, myocardial dysfunction, and heart failure), liver (*e.g.*, IR and nonalcoholic steatohepatitis) dysfunctions, CKD among other potential target organs with high-flow and low-resistance [36, 52].

6. Type 2 diabetes mellitus and cardiac autonomic neuropathy

Khandoker et al. conducted a comparative study of the characteristics of cardiac autonomic function alterations in patients with T2DM with diagnosed diabetic CKD, peripheral neuropathy (PN), and diabetic retinopathy. The results show that the entropy in patients, combining all complications, was significantly lower than the corresponding values for the control group. Odds ratios (OR) from entropy analysis also demonstrated a significantly higher association in patients with retinopathy and PN. Furthermore, the LF/HF ratio had a stronger connection with these diabetes-related complications, particularly in the group of patients who had all complications (OR: 4.92). The researchers suggest the type of microvascular or PN problem prevalent in T2DM persons affects HR entropy differently. In addition, attention is focused on implying that disorders of multi-organ connectivity are directly related to ANS dysfunction [42]. Barzilay et al. investigated whether measures of cardiovascular ANS function are linked with the incidence of diabetes and annual changes in FBG levels, as well as insulin sensitivity and secretion in older persons without diabetes. The mean annual unadjusted change in FBG was found to be 1 mg/dL. Higher detrended fluctuation analyses (DFA) values, averaged across 4–11 beats (DFA1), or 12–20 beats (DFA2), suggesting greater vs. less organization of beat-to-beat intervals, were related to reduced FBG increase with time. Higher SD of the N-N interval (SDNN) was related with decreased FBG rise with time in mutually adjusted analyses. Higher DFA1, DFA2, and SDNN levels were associated with increased insulin secretion and sensitivity but not with diabetes incidence. In individual and joint analyses, greater levels of specific cardiovascular autonomic factors linked with improved cardiovascular health are associated with a borderline decreased risk of diabetes incidents and significantly lower FBG level increases over time. These data support the concept that ANS function contributes to metabolic regulation [3].

7. Diabetic cardiac autonomic neuropathy and heart rate variability

Diabetic autonomic neuropathy, particularly CAN, is considered an important potential factor in circadian cardiovascular rhythms disruption [20, 53, 54]. The link between the symptoms of either greater sympathetic or decreased vagal activity and propensity for arrhythmogenesis has fueled efforts to establish quantitative autonomic activity markers [23, 55]. HRV is one of the most promising of these indicators [56]. The presence of CAN in diabetic patients is the strongest risk factor for early mortality and morbidity [20]. Landmark trials, such as the European Epidemiology and Prevention of Diabetes (EURODIAB) study, Action to Control Cardiovascular Risk in Diabetes (ACCORD) trial, and the Diabetes Control and Complications Trial/Epidemiology of Diabetes Interventions and Complications (DCCT/EDIC) study, have confirmed such an association. A cross-sectional investigation in T2DM conducted by Metelka et al. revealed severe autonomic dysfunction in the majority of patients studied, regardless of diabetes duration. It confirms the suggestion to test ANS integrity in T2DM patients at the time of diagnosis and during the early stages of the disease [57]. In this regard, the American Diabetes Association recommends screening for CAN in T2DM patients at the time of diagnosis [20].

A study conducted by Sethi et al. identified the prevalence of a very high proportion of CAN in T2DM, irrespective of the disease duration and glycemic control in asymptomatic patients. In addition, the obtained results indicate sympathetic and parasympathetic dysfunction, suggesting advanced CAN [58]. The results of the Verona newly diagnosed type 2 diabetes study showed that in 557 persons with newly onset T2DM, the prevalence of confirmed CAN was 1.8%, while the prevalence of early CAN was 15.3%. Therefore, it is likely that the pathophysiological disorders of metabolism are at an earlier stage, namely prediabetes [10].

Poor glycemic control in T1DM and a combination of HTN, obesity, DLP, and poor glycemic control in T2DM are established risk factors for CAN [20]. In patients with recent-onset diabetes, a lower vagus-mediated HRV has been shown to be associated with IR and reduced cardiorespiratory activity in both types of diabetes and hepatic steatosis [9] in T2DM. The obtained results indicate that these factors may contribute to the early development of CAN [59]. Kück et al. report that, unlike patients with new-onset T1DM, those with T2DM show early baroreflex dysfunction, likely due to IR and hyperglycemia [52]. Thus, dysglycemia is not the exclusive cause responsible for the initiation of CAN and its progression in T2DM. Obesity and its associated DLP, hyperinsulinemia, and HTN are additional risk factors for CAN in T2DM [44]. The Danish branch of the Anglo-Danish-Dutch Study of Intensive Treatment in People with Screen-Detected Diabetes in Primary Care (ADDITION-Denmark) aimed to investigate the course of CAN and related cardiometabolic risk variables in T2DM patients. Risk factors related to CAN status, as determined by cardiovascular autonomic reflex tests (CARTs), were studied using multivariate logistic regression. The results of the study showed a progressive heterogeneous course of CAN. Risk factors for CAN include hyperglycemia, obesity, and hypertriglyceridemia. Furthermore, hyperglycemia and obesity had a negative impact on continuous CAN measurements, highlighting the importance of modifiable risk factors in the development of CAN [27].

There are no established reference values for HRV variables that can be used to diagnose CAN [6]. However, Breder and Sposito suggest that CAN can be diagnosed by obtaining abnormal results in at least two of the six parameters listed below: SDNN 50 ms, the root mean square of the sum of squared differences (RMSSD) between R-R intervals <15 ms, the proportion of NN50 divided by the total number of NN (R-R) intervals (PNN50) < 0.75%, LF < 300 ms^2, and HF < 300 ms^2 derived from 24-hour Holter ECG recording [54]. Nighttime HRV may be a more accurate technique for measuring CAN and, as a result, may enhance the prediction of CV events in low-risk T2DM patients [60]. The Copenhagen Holter study recruited 678 community-dwelling individuals aged 55–75 years with no history of CVD. Six hundred fifty-three participants had access to both day and nighttime HRV. The study involved 133 participants with newly diagnosed T2DM and well-controlled T2DM. CV events were defined as CV death, myocardial infarction, stroke, or coronary revascularization. In persons with T2DM, 24-hour HRV was related to all-cause mortality rather than CV events. To predict CV events in persons with T2DM, conventional risk variables exhibited a receiver operating characteristic (ROC) value of 0.704 (95% CI 0.602–0.806). The addition of nighttime SDNN enhanced the prediction of CV events by conventional risk variables in persons with T2DM. Consequently, decreased nighttime HRV predicts an increased risk of CV events in adults with well-controlled T2DM, suggesting that nighttime HRV may supplement established risk variables in predicting CV events in T2DM patients [60].

Reduced HRV is the first indicator of CAN, indicating reduced parasympathetic and sympathetic activity without clinical signs and symptoms [28]. T2DM decreases practically every HRV variable. A systematic review and meta-analysis of 25 studies on HRV in T2DM revealed an overall decrease in HRV in T2DM persons due to a loss in both sympathetic and parasympathetic nerve function [53]. Another systematic review found that continuous RR intervals ratio (SD1/SD2), SDANN, and HF had higher sensitivity and specificity in detecting autonomic dysfunction in diabetes patients, suggesting they might be better diagnostic markers [24]. Abnormal nonlinear HRV variables are associated with diabetes or increase the risk of developing T2DM [55]. Similarly, a review study found lower HRV variables in MeTs and T2DM, as measured by short-term and 24-hour ECG recordings [28].

Pop-Busui et al. evaluated whether HRV measurements obtained from normal ECG recordings accurately assess CAN. Participants in the diabetes control and complications trial/epidemiology of diabetes interventions and complications underwent standardized CARTs (R-R response to paced breathing, Valsalva, and postural shifts), as well as digitized 12-lead resting ECGs. It is established that participants with CARTs-defined CAN had significantly lower SDNN and RMSSD compared with those without CAN (P < 0.001). SDNN dominates in defining CAN, with an area under the curve of 0.73 indicating acceptable test performance. For the best cutoff point, the Kappa statistic for SDNN was 0.41 (95% confidence interval 0.30–0.51), indicating good agreement with CARTs-defined CAN. These are the first studies to show convergence between HRV indices derived from ECGs and the gold standard CARTs, suggesting that they could measure CAN in clinical studies and therapeutic care [61].

8. Possible pathogenic pathways binding CAN and atherosclerosis progression

The link between CAN and atherosclerosis is widely established [46]. The ANS controls heart rate and vascular tone, and its failure may lead to atherosclerosis and arterial stiffness in diabetics [62]. According to the SEARCH cardiovascular disease project, even young patients with T1DM may show evidence of early autonomic dysfunction. Similarly, in these participants, CAN and arterial stiffness were linked independently of other traditional CV risk variables. However, whether CAN is related to simultaneous asymptomatic PAD in patients with arterial stiffness is unknown. Nattero-Chávez et al. postulated that CAN was linked to both arterial stiffness as measured by an ankle-branchial index (ABI) greater than 1.2 and the presence of PAD [45].

Nattero-Chávez et al. investigated the relationship between CAN and arterial stiffness as indicated by an ABI $\geq$ 1.2 in T1DM patients while thoroughly screening for PAD with vascular sonography. The authors present evidence that peripheral artery compliance is associated with cardiovascular autonomic dysfunction in young adults with T1DM who maintain satisfactory glycemic control.

The results reported by the author additionally demonstrate that the prevalence of CAN is threefold higher in patients with arterial stiffness than in those with normal ABI values, with the highest incidence in the group of patients with both concomitant PAD and arterial stiffness; this connection maintained even after adjusting for the presence of PAD or other relevant CV risk factors.

Furthermore, peripheral arterial stiffness encourages the relationship between cardiovascular autonomic dysautonomia and atherosclerosis in a group of T1DM patients. Other evidence supports this bidirectional pathogenic pathway from cardiac autonomic dysfunction to arterial stiffness and atherosclerosis [45].

Autonomic neuropathy appears to be more than just a microvascular consequence, with various pathophysiological mechanisms involved in its development [46].

9. The interplay between arterial stiffness and cardiac autonomic neuropathy

Diabetes is associated with two types of autonomic dysfunction: intrinsic and extrinsic [31]. The first is caused by a direct insult to the autonomic nerves, whereas the second can be induced by cardiovascular dysfunction, such as aortic stiffness and dilated cardiomyopathy. Studies on the principal causes of cardiac autonomic dysfunction in T2DM have revealed that it is primarily intrinsic [14, 44].

Diabetes mellitus is characterized by chronic hyperglycemia and in T2DM by IR, that is, chronic hyperinsulinemia. Dysfunction of the endothelial system is one of the first vascular complications observed in diabetic patients. Endothelial dysfunction worsens faster in T2DM patients than in T1DM patients, most likely due to the toxic effect of hyperinsulinemia on the arterial wall.

Endothelial dysfunction causes increasing arterial stiffness over time [16, 51]. Endothelial function and arterial elasticity decline with age, and the effect is exacerbated when diabetes is present [50].

The bidirectional relationship between OS and low-grade inflammation is also responsible for the deterioration of arterial structure and function in aging diabetes arteries. The primary molecule, nuclear factor kappa-light-chain-enhancer of activated B cells, is indirectly impacted by chronic hyperglycemia *via* ROS [63]. This specific molecule enables a positive feedback relationship between aging and diabetes. Diabetes causes further damage to aged arteries by downregulating antioxidant and anti-inflammatory mechanisms.

OS and inflammation may be engaged differently in diabetes than in arterial aging, with IR activating them in diabetes and a defective genetic longevity pathway in arterial aging [50].

As a result, diabetes affects all of the aging mechanisms discussed above. It has been proposed that the effects of diabetes and aging (a) share the same mechanisms, and thus cause additional damage to the arterial wall or (b) that they act in concert to amplify the deterioration of arterial structure and function caused by aging, in addition to its deleterious mechanisms that directly affect the arterial wall. The second theory is more correct based on the available data [50].

As a result, in middle-aged or older persons, arterial aging acts as the primary arterial wall failure, and diabetes acts as the secondary arterial wall failure seeded in the aging milieu. In other words, basic cellular dysfunction is caused by aging and senescence, exacerbated by secondary damage induced by IR and hyperglycemia in diabetes, resulting in secondary cell dysfunction.

The scenario is reversed in younger patients. Diabetes causes primary cellular dysfunction, whereas aging causes secondary cellular dysfunction. Although it may appear arbitrary, these two circumstances most likely overlap extensively. Nonetheless, these assumptions enhance the likelihood of arterial wall failure related to aging in diabetic patients [50].

Youth with T2DM have accelerated vascular aging, making them vulnerable to cardiovascular problems in early adulthood. Jaiswal et al. investigated the link between increased arterial stiffness and decreased HRV in young people with T2DM. In the SEARCH study, which enrolled 193 youth with T2DM were assessed PWV (PWV carotid-femoral segment) and HRV parameters. It was found that the youth with increased arterial stiffness were older, had higher BP, BMI, and TG, and lower HDL-C. In linear regression analysis, increased PWV was associated with lower SDNN independent of age, gender, BP, and BMI. However, when TG was taken into account, the correlation was diminished and nonsignificant, implying that the interaction between arterial stiffness and HRV is partly mediated by DLP [64].

Univariate analysis of the results of a cross-sectional study of 26 patients with DM revealed a significant positive correlation between resting systolic BP (SBP) and Ewing's score and an inverse correlation between the highest peak of volumes of O2 (VO2peak) and Ewing's score. Multivariate linear regression found that a significant model incorporating resting SBP and VO2peak explained 93.8% of the variance in Ewing's score. The author concluded that both factors were independent predictors of CAN in people with T2DM [62]. Thus, the current investigation demonstrated a moderate connection between high SBP and CAN, consistent with many prior studies. This link can be explained simply by the role of sympathetic activity dominance over parasympathetic activity in elevated BP.

The primary purpose of the ANS is to keep both arms of the ANS in balance. However, the equilibrium between the two arms is absent or decreased with CAN. This is indicated mainly by decreased parasympathetic activity and a dominating sympathetic impact over the heart and circulatory system, particularly the muscle tissue. Actually, sympathetic hyperactivity and higher sympathetic neural discharge are more prevalent in T2DM patients [65]. Moreover, *in vitro* investigations revealed that catecholamines (*e.g.*, adrenalin and noradrenaline) promote vascular smooth muscle replication, which may lead to vascular wall hyperplasia and arterial stiffening [62].

Eleftheriadou et al.'s cross-sectional study's major goal was to evaluate the association between PP amplification (PPA), HRV, and BRS in T2DM persons. Furthermore, the authors investigated the relationship between cardiac autonomic activity and central hemodynamic parameters influencing PPA such as aortic PWV, AIx, and common carotid artery stiffness distensibility coefficient. After correcting for age, duration of diabetes, height, waist circumference, aortic PWV, usage of β-blockers, and BRS-PPA was found to be substantially and independently linked with male gender, aortic SBP, HR, AIx, and HRV characteristics, such as total power of HRV.

No significant relationships were identified between HRV parameters or BRS and aortic PWV, AIx, or distensibility coefficient. Cardiovascular autonomic dysfunction was linked to increased PPA in T2DM patients. This connection was independent of the well-known influence of resting HR and standard CV risk or diabetes-related variables. Furthermore, it was not mediated by autonomic dysfunction's effects on arterial stiffness or pressure wave reflections. These results imply cardiac autonomic dysfunction influences PPA *via* mechanisms other than resting tachycardia and arterial characteristics [41].

What are the possible processes that connect cardio-autonomic dysfunction, arterial stiffness, and atherosclerosis? Complex interactions control the physiologic equilibrium of peripheral vascular beds and heart autonomic innervation, including metabolic processes such as OS and inflammation, which are disrupted in diabetic patients. Although CAN can cause an inflammatory response, other mechanisms, such as BP dysregulation, might play a role in developing AS and atherosclerosis [46, 47, 51]. From the early stages of CAN, parasympathetic downregulation changes the physiologic decline in BP at nighttime (*i.e.*, a so-called non-dipping pattern). In people with CAN, the relative sympathetic overload and exposure of the vascular bed to elevated BP values during sleep may, at least in theory, induce vascular damage, stiffening of the arteries, and atherosclerosis [45].

Alternatively, from the early stages of T1DM, autonomic dysfunction and a decrease in arterial elasticity may coexist [61]; however, their order of appearance and the potential involvement of causality in such an association are unsolved. Nattero-Chávez et al. indicated that the coexistence of arterial stiffness and PAD partly explained the association between CAN and arterial stiffness. The cross-sectional design, however, precludes any conclusions concerning causality [45].

In theory, arterial stiffness may lead to cardiovascular dysautonomia due to baroreceptor dysfunction; conversely, CAN may encourage arterial stiffness by increasing HR as an increase in HR results in arterial stiffening regardless of ANS activity changes [10]. Furthermore, cardiovascular dysautonomia may affect arterial wall elasticity by changing the vascular tone of large arteries [45].

CAN and arterial stiffness may develop in parallel due to aging in hyperglycemia rather than being directly connected. Hyperglycemia leads to atherosclerosis *via* various pathways, including endothelial dysfunction and hypercoagulability [48]. Chronic hyperglycemia enhances the accumulation of AGEs, which disrupt the adhesion capabilities of endothelial cells' basement membranes and activate inflammatory cells in the arterial wall, favoring atherogenesis [48]. Similarly, AGEs cause collagen cross-linking within the vascular wall, losing collagen elasticity, and decreasing arterial and cardiac compliance [45].

10. Conclusion

CAN is one of the underdiagnosed microvascular complications of T2DM caused by hyperglycemia-induced neuronal damage [54]. The decline in HRV is seen even before manifesting signs and symptoms of diabetic CAN. As a result, there is strong evidence of decreased HRV in T2DM patients. HRV pattern analysis has the capacity to discover autonomic imbalance in the preclinical and asymptomatic stages. Both sympathetic and parasympathetic activity is reduced in T2DM persons, which can be explained by the negative effects of altered glucose metabolism on HRV [53, 54].

Arterial stiffness may be involved in developing several dysfunctions in the heart, brain, liver, kidney, and others. Arterial stiffness may contribute to cardiovascular dysautonomia by inducing baroreceptor dysfunction; conversely, CAN may favor arterial stiffness by increasing HR and impairing arterial wall elasticity. Both states may also develop in parallel due to aging in the presence of hyperglycemia.

The presence of CAN should be evaluated considerably earlier in the DM process, and reduced HRV is the earliest sign of CAN. Improvement of HRV may allow guiding the patients toward lifestyle changes and early management.

Given its promise as a noninvasive, reliable, and painless assessment, the benefits of an HRV evaluation in diagnosing and monitoring the severity of T2DM should be investigated further.

Conflict of interest

The authors declare no conflict of interest.

Author details

Victoria Serhiyenko[1,3]*, Marta Hotsko[1], Yuriy Markevich[2], Martyn-Yurii Markevich[2], Volodymyr Segin[3], Ludmila Serhiyenko[4] and Alexandr Serhiyenko[1]

1 Department of Endocrinology, Danylo Halytsky Lviv National Medical University, Lviv, Ukraine

2 Lviv Regional Clinical Hospital, Lviv, Ukraine

3 Medical Center "Symbiotyka Cardio", Lviv, Ukraine

4 Department of Medical Biology, Danylo Halytsky Lviv National Medical University, Lviv, Ukraine

*Address all correspondence to: serhiyenkoa@gmail.com

References

[1] Serin Y, Acar TN. Effect of circadian rhythm on metabolic processes and the regulation of energy balance. Annals of Nutrition & Metabolism. 2019;**74**(4):322-330

[2] Russo B, Menduni M, Borboni P, Picconi F, Frontoni S. Autonomic nervous system in obesity and insulin-resistance-the complex interplay between leptin and central nervous system. International Journal of Molecular Sciences. 2021;**22**(10):5187

[3] Barzilay JI, Tressel W, Biggs ML, et al. The association of measures of cardiovascular autonomic function, heart rate, and orthostatic hypotension with incident glucose disorders: The cardiovascular health study. Diabetes Care. 2022;**45**(10):2376-2382

[4] Faber CL, Deem JD, Campos CA, Taborsky GJ Jr, Morton GJ. CNS control of the endocrine pancreas. Diabetologia. 2020;**63**(10):2086-2094

[5] Ulleryd MA, Prahl U, Börsbo J, et al. The association between autonomic dysfunction, inflammation and atherosclerosis in men under investigation for carotid plaques. PLoS One. 2017;**12**(4):e0174974

[6] Serhiyenko V, Serhiyenko A. Diabetic cardiac autonomic neuropathy. In: Rodriguez-Saldana J, editor. The Diabetes Textbook. 2nd ed. Basel: Springer. Cham. Springer Nature Switzerland AG; 2023. pp. 939-966

[7] Spallone V, Ziegler D, Freeman R, et al. Cardiovascular autonomic neuropathy in diabetes: Clinical impact, assessment, diagnosis, and management. Diabetes/ Metabolism Research and Reviews. 2011;**27**(7):639-653

[8] Pop-Busui R, Boulton AJM, Feldman EL, et al. Diabetic neuropathy: A position statement by the American Diabetes Association. Diabetes Care. 2017;**40**(1):136-154

[9] Ziegler D, Porta M, Papanas N, et al. The role of biofactors in diabetic microvascular complications. Current Diabetes Reviews. 2022;**18**(4):e25082115830

[10] Serhiyenko VA, Serhiyenko LM, Serhiyenko AA. Features of circadian rhythms of heart rate variability, arterial stiffness and outpatient monitoring of blood pressure in diabetes mellitus: Data, mechanisms and consequences. In: Sinha RP, editor. Circadian Rhythms and Their Importance. New York: Nova Science Publishers; 2022. pp. 279-341

[11] Aroor AR, Jia G, Sowers JR. Cellular mechanisms underlying obesity-induced arterial stiffness. American Journal of Physiology-Regulatory, Integrative and Comparative Physiology. 2018;**314**(3):R387-R3R8

[12] Chorepsima S, Eleftheriadou I, Tentolouris A, et al. Pulse wave velocity and cardiac autonomic function in type 2 diabetes mellitus. BMC Endocrine Disorders. 2017;**17**(1):27

[13] Clyburn C, Sepe JJ, Habecker BA. What gets on the nerves of cardiac patients? Pathophysiological changes in cardiac innervation. The Journal of Physiology. 2022;**600**(3):451-461

[14] Cseh D, Climie RE, Offredo L, et al. Type 2 diabetes mellitus is independently associated with decreased neural baroreflex sensitivity: The Paris prospective study III. Arteriosclerosis,

Thrombosis, and Vascular Biology. 2020;**40**(5):1420-1428

[15] Zilliox LA, Russell JW. Is there cardiac autonomic neuropathy in prediabetes? Autonomic Neuroscience. 2020;**229**:102722

[16] Liu L, Wu Q, Yan H, Chen B, Zheng X, Zhou Q. Association between cardiac autonomic neuropathy and coronary artery lesions in patients with type 2 diabetes. Disease Markers. 2020;**2020**:6659166

[17] Li Y, Liu Y, Liu S, et al. Diabetic vascular diseases: Molecular mechanisms and therapeutic strategies. Signal Transduction and Targeted Therapy. 2023;**8**(1):152

[18] Bhati P, Alam R, Moiz JA, Hussain ME. Subclinical inflammation and endothelial dysfunction are linked to cardiac autonomic neuropathy in type 2 diabetes. Journal of Diabetes and Metabolic Disorders. 2019;**18**(2):419-428

[19] Chen Y, Yu Y, Zou W, Zhang M, Wang Y, Gu Y. Association between cardiac autonomic nervous dysfunction and the severity of coronary lesions in patients with stable coronary artery disease. The Journal of International Medical Research. 2018;**46**(9):3729-3740

[20] Spallone V. Update on the impact, diagnosis and management of cardiovascular autonomic neuropathy in diabetes: What is defined, what is new, and what is unmet. Diabetes and Metabolism Journal. 2019;**43**:3-30

[21] Williams S, Raheim SA, Khan MI, et al. Cardiac autonomic neuropathy in type 1 and 2 diabetes: Epidemiology, pathophysiology, and management. Clinical Therapeutics. 2022;**44**(10):1394-1416

[22] Addepalli V, Suryavanshi SV. Catechin attenuates diabetic autonomic neuropathy in streptozotocin induced diabetic rats. Biomedicine & Pharmacotherapy. 2018;**108**:1517-1523

[23] Vigo DE, Siri LN, Cardinali DP. Heart rate variability: A tool to explore autonomic nervous system activity in health and disease. In: Gargiulo P, Mesones Arroyo H, editors. Psychiatry and Neuroscience Update. Berlin/Heidelberg: Springer Nature Switzerland; 2019. pp. 113-126

[24] França da Silva AK, Penachini da Costa de Rezende Barbosa M, Marques Vanderlei F, Destro Christofaro DG, Marques Vanderlei LC. Application of heart rate variability in diagnosis and prognosis of individuals with diabetes mellitus: Systematic review. Annals of Noninvasive Electrocardiology. 2016;**21**(3):223-235

[25] Black N, D'Souza A, Wang Y, et al. Circadian rhythm of cardiac electrophysiology, arrhythmogenesis, and the underlying mechanisms. Heart Rhythm. 2019;**16**(2):298-307

[26] Malik M, Hnatkova K, Huikuri HV, et al. CrossTalk proposal: Heart rate variability is a valid measure of cardiac autonomic responsiveness. The Journal of Physiology. 2019;**597**(10):2595-2598

[27] Andersen ST, Witte DR, Fleischer J, et al. Risk factors for the presence and progression of cardiovascular autonomic neuropathy in type 2 diabetes: ADDITION-Denmark. Diabetes Care. 2018;**41**(12):2586-2594

[28] Trivedi GY, Saboo B, Singh RB, Maheshwari A, Sharma K, Verma N. Can decreased heart rate variability be a marker of autonomic dysfunction, metabolic syndrome and diabetes? Journal of Diabetology. 2019;**10**(2):48-56

[29] Yeh SJ, Lung CW, Jan YK, Liau BY. Advanced cross-correlation function application to identify arterial baroreflex sensitivity variations from healthy to diabetes mellitus. Frontiers in Neuroscience. 2022;**16**:812302

[30] Kaufmann H, Norcliffe-Kaufmann L, Palma JA. Baroreflex dysfunction. The New England Journal of Medicine. 2020;**382**(2):163-178

[31] Goldberger JJ, Arora R, Buckley U, Shivkumar K. Autonomic nervous system dysfunction: JACC focus seminar. Journal of the American College of Cardiology. 2019;**73**(10):1189-1206

[32] Dimova R, Chakarova N, Grozeva G, Tankova T. Evaluation of the relationship between cardiac autonomic function and glucose variability and HOMA-IR in prediabetes. Diabetes & Vascular Disease Research. 2020;**17**(5):1479164120958619

[33] Moulle VS, Tremblay C, Castell AL, et al. The autonomic nervous system regulates pancreatic b-cell proliferation in adult male rats. American Journal of Physiology-Endocrinology and Metabolism. 2019;**317**:E234-EE43

[34] Miller BM, Oderberg IM, Goessling W. Hepatic nervous system in development, regeneration, and disease. Hepatology. 2021;**74**(6):3513-3522

[35] Durham AL, Speer MY, Scatena M, Giachelli CM, Shanahan CM. Role of smooth muscle cells in vascular calcification: Implications in atherosclerosis and arterial stiffness. Cardiovascular Research. 2018;**114**(4):590-600

[36] Chirinos JA, Segers P, Hughes T, Townsend R. Large-artery stiffness in health and disease: JACC state-of-the-art review. Journal of the American College of Cardiology. 2019;**74**(9):1237-1263

[37] Bissinger A. Cardiac autonomic neuropathy: Why should cardiologists care about that? Journal Diabetes Research. 2017;**2017**:5374176

[38] Çakar M, Balta Ş, Şarlak H, et al. Arterial stiffness and endothelial inflammation in prediabetes and newly diagnosed diabetes patients. Archives of Endocrinology and Metabolism. 2015;**59**(5):407-413

[39] Bianchi L, Chiheb S, Banu I, Rezki A, Cosson E. Influence of cardiac autonomic dysfunction and arterial stiffness on subendocardial myocardial viability in patients with type 2 diabetes. Diabetes & Metabolism. 2016;**42**(4):297-298

[40] Bagherzadeh A, Nejati-Afkham A, Tajallizade-Khoob Y, et al. Association of cardiac autonomic neuropathy with arterial stiffness in type 2 diabetes mellitus patients. Journal of Diabetes and Metabolic Disorders. 2013;**12**(1):55

[41] Eleftheriadou I, Drosos GC, Tentolouris A, et al. Pulse pressure amplification and cardiac autonomic dysfunction in patients with type 2 diabetes mellitus. Journal of Human Hypertension. 2018;**32**(8-9):531-539

[42] Khandoker AH, Al-Angari HM, Khalaf K, et al. Association of diabetes related complications with heart rate variability among a diabetic population in the UAE. PLoS One. 2017;**12**(1):e0168584

[43] Paneni F, Diaz Cañestro C, Libby P, Lüscher TF, Camici GG. The aging cardiovascular system: Understanding it at the cellular and clinical levels. Journal of the American College of Cardiology. 2017;**69**(15):1952-1967

[44] Bakkar NZ, Mougharbil N, Mroueh A, et al. Worsening baroreflex

sensitivity on progression to type 2 diabetes: Localized vs. systemic inflammation and role of antidiabetic therapy. American Journal of Physiology-Endocrinology and Metabolism. 2020;**319**(5):E835-EE51

[45] Nattero-Chávez L, Redondo López S, Alonso Díaz S, et al. Association of cardiovascular autonomic dysfunction with peripheral arterial stiffness in patients with type 1 diabetes. The Journal of Clinical Endocrinology and Metabolism. 2019;**104**(7):2675-2684

[46] Mala S, Hoskovcova L, Riedlbauchova L, Nedelka T, Broz J. Relationship between cardiac autonomic neuropathy and atherosclerosis in patients with diabetes mellitus. Current Research in Diabetes & Obesity Journal. 2018;**9**(1):555753

[47] Lacolley P, Regnault V, Laurent S. Mechanisms of arterial stiffening: From mechanotransduction to epigenetics. Arteriosclerosis, Thrombosis, and Vascular Biology. 2020;**40**(5):1055-1062

[48] Haas AV, McDonnell ME. Pathogenesis of cardiovascular disease in diabetes. Endocrinology and Metabolism Clinics of North America. 2018;**47**(1):51-63

[49] Zheng M, Zhang X, Chen S, et al. Arterial stiffness preceding diabetes: A longitudinal study. Circulation Research. 2020;**127**(12):1491-1498

[50] Lunder M, Janić M, Šabovič M. Treating arterial ageing in patients with diabetes: From mechanisms to effective drugs. International Journal of Molecular Sciences. 2021;**22**(6):2796

[51] Kozakova M, Palombo C. Diabetes mellitus, arterial wall, and cardiovascular risk assessment. International Journal of Environmental Research and Public Health. 2016;**13**(2):201

[52] Kück JL, Bönhof GJ, Strom A, et al. Impairment in baroreflex sensitivity in recent-onset type 2 diabetes without progression over 5 years. Diabetes. 2020;**69**(5):1011-1019

[53] Benichou T, Pereira B, Mermillod M, et al. Heart rate variability in type 2 diabetes mellitus: A systematic review and meta-analysis. PLoS One. 2018;**13**(4):e0195166

[54] Ferdousi S, Gyeltshen P. Type 2 diabetes mellitus: Cardiovascular autonomic neuropathy and heart rate variability. In: Pantea-Stoian A, editor. Type 2 Diabetes: From Pathophysiology to Cyber Systems. London, UK: IntechOpen; 2021. pp. 215-306

[55] Vinik AI, Casellini C, Parson HK, Colberg SR, Nevoret ML. Cardiac autonomic neuropathy in diabetes: A predictor of cardiometabolic events. Frontiers in Neuroscience. 2018;**12**:591

[56] Shah AS, El Ghormli L, Vajravelu ME, et al. Heart rate variability and cardiac autonomic dysfunction: Prevalence, risk factors, and relationship to arterial stiffness in the treatment options for type 2 diabetes in adolescents and youth (TODAY) study. Diabetes Care. 2019;**42**(11):2143-2150

[57] Metelka R, Cibičková L, Gajdová J, Krystyník O. Heart rate variability evaluation in the assessment of cardiac autonomic neuropathy in patients with type 2 diabetes. Cor et Vasa. 2018;**60**(4):e335-ee44

[58] Sethi PP, Jatteppanavar B, Kant R, Pathania M, Bairwa MC. Spectrum of cardiac autonomic neuropathy in patients with type 2 diabetes mellitus: A north

India perspective. Journal of Cardio-Diabetes and Metabolic Disorders. 2022;**2**(1):23-28

[59] Herder C, Schamarek I, Nowotny B, et al. Inflammatory markers are associated with cardiac autonomic dysfunction in recent-onset type 2 diabetes. Heart. 2017;**103**(1):63-70

[60] Hadad R, Larsen BS, Weber P, et al. Night-time heart rate variability identifies high-risk people among people with uncomplicated type 2 diabetes mellitus. Diabetic Medicine. 2021;**38**(7):e14559

[61] Pop-Busui R, Backlund JC, Bebu I, et al. Utility of using electrocardiogram measures of heart rate variability as a measure of cardiovascular autonomic neuropathy in type 1 diabetes patients. Journal of Diabetes Investigation. 2022;**13**(1):125-133

[62] Osailan A. Cardiovascular autonomic neuropathy in people with type 2 diabetes mellitus; investigation of its association with classical cardiovascular risk factors using cardiovascular autonomic reflex tests: A cross-sectional study. The Egyptian Heart Journal. 2021;**73**(1):44

[63] Assar ME, Angulo J, Rodríguez-Mañas L. Diabetes and ageing-induced vascular inflammation. The Journal of Physiology. 2016;**594**(8):2125-2146

[64] Jaiswal M, Divers J, Urbina EM, et al. Cardiovascular autonomic neuropathy in adolescents and young adults with type 1 and type 2 diabetes: The SEARCH for diabetes in youth cohort study. Pediatric Diabetes. 2018;**19**(4):680-689

[65] López-Cano C, Gutiérrez-Carrasquilla L, et al. Sympathetic hyperactivity and sleep disorders in individuals with type 2 diabetes. Frontiers in Endocrinology (Lausanne). 2019;**10**:752

Chapter 8

Clinical, Pathophysiological and Electrodiagnostic Aspects of Lambert-Eaton Myasthenic Syndrome

Felipe Fanine de Souza, Julia Petry Trevisani and Felipe Ibiapina dos Reis

Abstract

Lambert-Eaton myasthenic syndrome (LEMS) is characterized by an autoimmune disorder of the neuromuscular junction, which, through a reduction in nerve terminal acetylcholine release mediated by antibodies against functional voltage-gated calcium channels (VGCCs) of the P/Q in presynaptic nerve terminals, leads to proximal muscle weakness, in addition to autonomic dysfunction and areflexia, constituting the classic triad of symptoms. The syndrome presents itself in two forms: the paraneoplastic form—resulting mainly from small cell lung carcinoma—and the underlying autoimmune form. With clinical suspicion of the disease, the diagnosis can be made through serological and electrophysiological tests, which present typical findings and reflect the existence of a presynaptic transmission defect. Treatment is based on early screening and removal of the etiological agent, which in the most common case is the treatment of the underlying cancer. In patients whose symptoms affect their daily activities, some medications can intervene in the search for a better quality of life, such as amifampridine, pyridostigmine and 3,4-diaminopyridine (3,4-DAP). It must be remembered, however, that LEMS has a significant impact on the patient's quality of life and ability to perform daily activities and therefore warrants timely diagnosis and adequate treatment in itself.

Keywords: Lambert-Eaton, neuromuscular junction, miasthenic, acetylcholine, paraneoplastic, autoantibodies, cancer

1. Introduction

Lambert-Eaton myasthenic syndrome (LEMS) is characterized as an uncommon disorder of the neuromuscular junction, in which a considerable percentage of cases present as a paraneoplastic form (P-LEMS) associated, in most cases, with lung carcinoma of small cells (CPPC). Another portion of the disorders is associated with underlying autoimmune pathologies (A-LEMS) [1].

LEMS is characterized by the presence of antibodies against presynaptic voltage-gated calcium channels (VGCCs) of the P/Q type, which provide a decrease in the levels of acetylcholine (ACh) that are released in the nerve terminal, providing one of the main symptoms encountered, such as weakness and fatigue. Some clinical symptoms of LEMS overlap with those of other myasthenic syndromes, most commonly myasthenia gravis (MG), which can contribute to misdiagnosis or delay in diagnosis [1].

In addition, LEMS is a disorder in which the pathophysiological, clinical, electrophysiological and laboratory characteristics are distinct, with the presence of common clinical findings among other pathologies, such as autonomic dysfunction and areflexia [1].

Despite being an uncommon disorder in a population sample, among the conditions involving neuromuscular transmission, knowledge about the disease is common, in which the clinical presentation, mainly associated with neoplasia, requires the neurologist to have knowledge about its presentation, diagnosis and treatment [2].

The prognosis of the disease is related to the presence of cancer and the severity/distribution of muscle weakness, in which the cause of death in these patients is mainly due to tumor progression. LEMS is a clinically important early indicator of possible cancer; therefore, a diagnosis of LEMS should immediately prompt rigorous oncologic screening and surveillance [1, 2].

2. Material and methods

This respective chapter is a non-systematic review. All articles used were searched in PubMed, Medline, ScienceDirect, SciELO and Cochrane public databases. The search process for these articles was carried out by searching for terms related to Lambert-Eaton syndrome, selecting the most relevant articles available in English or Portuguese. Additional literature not present in these databases was used to complement the general understanding of the conditions, specific classifications, previous treatments and also to illustrate historical descriptions of the disease.

3. Epidemiology

LEMS is considered a rare disease, which is difficult to diagnose and somewhat underdiagnosed, with an annual incidence corresponding to one-tenth of cases of MG—the main disease in the group of neuromuscular junction disorders [3]. Sanders [4], in a large North American epidemiological study, estimated that the prevalence was 1 in 100,000 of confirmed and probable cases, respectively, being more frequently presented in males and in the older population.

When we refer to the approximate duration of disease prior to the diagnosis of the case, we have approximately 11 months and it was significantly lower among patients with paraneoplastic etiology than those with the autoimmune portion [5]. In addition, in P-LEMS, the average age of onset was 60 years, with 65–75% of patients being male, as previously mentioned [6].

Elrington et al., in a prospective study, demonstrated that LEMS has a higher incidence in males, ranging from 60% to 75% of patients, which differs from MG, where most of the conditions present in women. Another factor to be noted is the age of onset, in which the patient with the presence of a non-paraneoplastic form, the age is similar to that of MG, ranging around 35 years. In contrast, in paraneoplastic LEMS, the peak incidence generally remains at 58 years of age. When we compare the prevalence of MG compared to LEMS, the occurrence of the first disease is 46 times higher.

4. Immunophysiology

Under physiological conditions, it would be normal for the depolarization of the presynaptic neuronal membrane to induce the opening of voltage-gated calcium channels (VGCC), which would cause the influx of calcium to the nerve terminal, which would act in muscle contraction. Through the neuromuscular junction. ACh—neurotransmitter active in this process—diffuses through the synapse to bind to its receptors on the postsynaptic membrane of the motor endplate, and this binding opens postsynaptic sodium and potassium channels, promoting depolarization of the endplate motor. After reaching the depolarization threshold, there is an action potential and muscle contraction [7].

Calcium ions play a key role in neurotransmission, where they not only play a role in the performance of vesicular exocytosis that are anchored on the cytoplasmic side of the presynaptic membrane (ready to be released), but also play a role in short-term synaptic plasticity and, probably influence the mechanisms that restore the RRP after presynaptic activation [7].

Figure 1 succinctly illustrates the most relevant stages of neuromuscular transmission, in which first (1) we have the active depolarization of a motor axon promoted by the opening of Na^+ channels and, in addition, the membrane contains dependent K^+ and Ca^{+2} channels voltage (2). The opening of calcium channels (3) located in the active zones allows the entrance of this, generating an increase in the intra-concentration. Approximately four or five calcium ions bind to its receptor, which triggers the rapid and synchronous release of ACh (4). Subsequently, the empty active zones are replenished by a recycling pool®, which is believed to be mediated by the intraterminal concentration of calcium ions (5).

In Lambert-Eaton syndrome, the body produces autoantibodies to the presynaptic VGCC, which therefore results in a reduction of Ach released in the presynaptic nerve terminal [8]. Given this, by altering the release of Ach, the entire cascade of neurotransmission and channel opening mentioned above end up reducing as well. A decreased amount of Ach translates into an underactivation of sodium and potassium channels in the postsynaptic membrane and reduces the action potential in the endplate [7]. Therefore, the action potential ends up not occurring and, consequently, affects muscle contraction, which explains one of the main symptoms of the disease, muscle weakness.

4.1 Tumor association

Association with tumor is reported in about 60% of patients with LEMS [9], in which the majority is due to smoking-related SCLC with neuroendocrine characteristics, but other malignant neoplasms are described in the literature, such as small cell lung carcinomas. Non-small and mixed, prostate carcinoma, thymoma and lymphoproliferative disorders [10, 11]. Studies indicate that the diagnosis of LEMS can precede the diagnosis of neoplasia by a variability of 5–6, in addition to having an extreme relationship with the smoker, which is presented as a risk factor. Titulaer et al. [16] observed that an SCLC diagnosis preceded LEMS identification in only 6% of P-LEMS cases. SCLCs were identified in 92% of these patients within 3 months and in 96% within a year.

Descriptions in the literature show assumptions about the initial human autoimmune response in patients with LEMS being generated against the antigens of the VGCC subunit in lung carcinoma [12], and the paraneoplastic form expresses VGCC of the N, L or P type, in addition to the P/Q subtypes and possibly N are targets of IgG-mediated autoimmunity in LEMS [13] (P/Q are involved in Ach release from motor nerve terminals and N-type in peripheral autonomic terminals [14]).

5. Clinical manifestations

The clinical presentation of LEMS is variable and has an insidious onset, with a slowly progressive clinical course, but occasionally it can be subacute. Presenting symptoms are lower limb weakness (60%) or generalized weakness (18%), muscle pain or stiffness (5%), dry mouth (5%), upper limb weakness (4%), diplopia (4%) and dysarthria (2%). In this aspect, there is a classic triad that can further characterize the disease, with proximal muscle weakness, autonomic characteristics and areflexia [9, 11].

Titulaer et al. [15] (**Table 1**) described the spread of weakness in patients with A-LEMS and P-LEMS, which is more characterized by spreading from proximal to distal, affecting feet and hands, from caudal to cranial and, finally, reaching the oculobulbar region. Analyzing and comparing with MG, LEMS presents a different presentation. In addition, the speed of propagation is more pronounced in the paraneoplastic form, in which generalized weakness can already be noticed in the first 3 months [16].

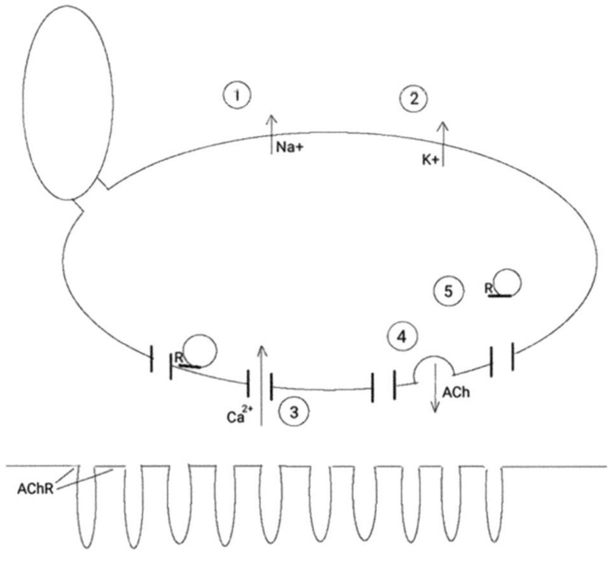

Figure 1.
Stages of neuromuscular transmission.

	Lambert-Eaton	Myasthenia Gravis
Typical first symptom	Difficulty getting up	Diplopia
Autonomic symptom	Dry mouth	None
Weaker musculature	Proximal muscle weakness	Extraocular
Deep tendon reflexes	Reduced	Normal
Involvement of autoantibodies	Presynaptic VGCC	Postsynaptic ACh receptor
High frequency RNS	>60% increase in muscle action potential	<60%

Table 1.
Comparison of symptoms and characteristics of LEMS and MG.

In the study conducted by Burns et al. [17], oculobulbar symptoms/signs were present in 78% of patients with the disease. In addition, another retrospective review at the Mayo Clinic in Minnesota evaluated 176 patients, who described the following findings: 23% of patients had ptosis, 20.5% diplopia, 14% decreased visual acuity, in addition to 8, 5% alteration in the extrinsic ocular musculature. In a general context, oculobulbar symptoms present in a more severe and prolonged course of the disease [18].

Autonomic dysfunction is reported in 80–96% of patients, and the most common complaint is dry mouth, in addition to other symptoms such as erectile dysfunction, constipation, orthostatic dysfunction and difficulty urinating [15, 16].

Decreased tendon reflexes or areflexia are common signs we find, which should be tested after a period of rest, as the phenomenon of post-exercise facilitation can mask this characteristic finding since they can be amplified after muscle contraction. Cerebellar ataxia, sensory neuropathy and limbic encephalitis are extremely uncommon findings that are almost completely associated with the paraneoplastic form [16, 19].

6. Diagnostic

As previously seen, LEMS is initially suspected based on the patient's clinical condition, in which the presence of clinical signs and symptoms should be noted, demonstrating the classic triad of proximal muscle weakness, diminished tendon reflexes and autonomic dysfunction. Based on this principle and having a clinical basis for the diagnosis, there are different electrophysiological studies, such as electromyography (EMG) and nerve conduction studies (NCS) that can contribute to confirming the diagnostic doubt [15].

The diagnosis of LEMS is by combining clinical suspicion with confirmation through electrophysiology and antibody testing against the P/Q-type VGCC IgG antibody (VGCC). The first analysis is through EMG that confirms the presence of a presynaptic disorder, which the physiopathology will be explained in "Electrophysiology" chapter [1, 2].

The importance of antibody research is to predict cancer development, due to half of LEMS patients having associated SCLC15. Nearly 90% of LEMS cases are seropositive for P/Q- type VGCC antibodies, once the disease acts impairing the release of ACh at active zones [20].

Besides, there are several different neuronal VGCC subtypes that also remained highly sensitive and specific for SCLC in all LEMS patients compared to NT-LEMS collaborating for early diagnosis, as N-type VGCCs (reportedly in 33–49% of LEMS)1, anti-Sry-like High Mobility Group Box (SOX), HuD and GABAb [21].

In previous reports, it was found that the presence of either N-type VGCC, SOX2 or GABAb antibodies was 84% sensitive and 87% specific for the SCLC detection [22]. Antibodies against SOX are considered malignant neoplasm-related onconeural autoabs, due to the association with SCLC. Sun et al. found among the 110 anti-SOX1-abs-positive, LEMS (30.0%, n = 33) was the most common paraneoplastic neurological syndrome in patients with cancer.

Autoimmunity with SCLC: In patients with LEMS associated with SCLC, the tumor tissue expresses VGCC, which induces the production of autoantibodies, which ultimately cross-react with presynaptic VGCC antigens [23].

Genetic predisposition: Non-tumor LEMS is more associated with class I HLA-B8 and class II HLA-DR3 and DQ2. These genotypes have been associated with other autoimmunity conditions, which include MG. However, this association does not appear when we have LEMS due to SCLC [23].

6.1 Screening for malignancy

Because of the high association with malignancy, the diagnosis of the myasthenic syndrome should lead to immediate and extensive search and screening for underlying malignant processes. Among them, computed tomography (CT) of the chest or magnetic resonance imaging (MRI) would be the recommended initial imaging study. PET scan can also be used in this evaluation if the first studies are negative. In the face of a first negative analysis, screening should continue and be performed every 3–6 months over a period of at least 2 years, in which case a quarterly assessment should be indicated for those at high risk, whose DELTA-P score is greater than 2 or with positive SOX antibodies associated with SCLC-LEMS [23, 24].

DELTA-P is a simple clinical scoring system based on age, weight loss, smoking, bulbar involvement, erectile dysfunction, and Karnofsky performance status, called the LEMS Tumor Association Prediction Score, which was developed and evaluated in 2011. This scoring system indicates the presence of an SCLC with very high accuracy and therefore helps clinicians to identify high-risk patients and optimize the triage and follow-up process [6].

6.2 Electrophysiology

Among the established criteria and the electrophysiology of LEMS, repetitive nerve stimulation (RNS) is the test of choice, making it possible to analyze characteristic findings that are part of the classic triad in the diagnosis, such as:

1. Decreased compound motor action potentials (CMAPs) in resting NCS (reaching less than 50% of the lower limit of normal), which is a common finding in all disorders of the presynaptic terminal of the neuromuscular junction, in addition, to be observed in approximately 96% of cases of LEMS [23].

2. Decrease in CMAP response at low frequency (2–5 Hz), which produces a successive decline in CMAP amplitude from its normal baseline. A decrease greater

than 10% is considered abnormal and this presentation is found in approximately 94–98% of patients with LEMS [23, 24].

3. Increase in response greater than 100% demonstrated immediately after 10–30 seconds of maximum voluntary contraction (post-exercise test), or with high-frequency stimulation (20–50 Hz)—an incremental response has been used as the electrophysiological gold standard for the diagnosis of LEMS [23, 24].

Needle EMG shows erratic changes in motor unit action potential such as low and short during voluntary action potential. This may be due to single-fiber EMG measurements on the jitter, which increased jittering shows transmission blockage and may correspond to disease severity [1, 23, 24].

Fiber EMG usually does not differ between presynaptic and postsynaptic disorders of the neuromuscular junction. It should be noted that early in the clinical course of LEMS, when the initial CMAP amplitude is not yet reduced, the abnormal incremental response may be easier to demonstrate than the more characteristic incremental response. Given this, the combination of a normal CMAP or when it is slightly reduced, in addition to a lack of incremental response, can lead to a misdiagnosis of LEMS or lead to a misdiagnosis of MG[1].

Hatanaka and Oh [25], in a study conducted in 2008, showed that longer exercises can lead to a progressive decrease in the increment, in which the diagnostic sensitivity is greater using a 10-second exercise when compared to a 30-second exercise. The same authors found in later studies that, when comparing negative LEMS VGCC, the positive result of this serology is associated with a lower baseline CMAP and a greater response to the increment. These findings suggest that the RNS pattern is more drastically altered in positive VGCC and thus becomes more indicative of LEMS. Therefore, this study contributed even more with the use of the 60% increment criterion, which is quite critical for the diagnosis of seronegative LEMS. Finally, the authors also conclude that the effect of exhaustion after exercise on the RNS has no diagnostic value [26].

7. Treatment

The treatment itself for the onset of the disease is based on the underlying etiology, requiring monitoring and treatment of the cancer, if it has a neoplastic etiology. However, there are drugs that can act directly on the synaptic terminal, improving quality of life and reducing symptoms. Among them, we have amifampridine, which is an oral medication that acts on voltage-dependent potassium channels in the presynaptic terminal, blocking them, thus preventing the efflux of potassium ions and prolonging depolarization. In this mechanism, there is an increase in the release of Ach through the prolongation of the influx of calcium ions, which will promote an improvement in the neuromuscular function. This medication has been approved in the USA as intended to be first-line therapy in the treatment of the syndrome by the Food and Drug Administration (FDA) (**Table 2**) [27].

Another medication that can act on the mechanism of Ach release and promote the improvement of neuromuscular function is pyridostigmine, which inhibits acetylcholinesterase and, consequently, increases its concentration in the motor plate synapse [28]. However, this medication is generally not very effective in

Lambert-Eaton myasthenic syndrome—approach to treatment	
Screen for small cell cancer	If cancer is detected then aggressive treatment for malignancy may improve or be curative for LEMS
Amifampridine	Begin 5 mg three times per day and gradually increase every 3–4 days by 5 mg
Pyridostigmine	30–60 mg three to four times per day
If insufficient improvement then add immunosuppressant therapy	Prednisone, azathioprine, mycophenolate, etc.
If severe or refractory weakness the add IVIg/PE Rituximab	IVIg 2 g/kg over 4–5 days then maintenance or plasma exchange

Table 2.
Options for treatment of LEMS include treating the malignancy (if paraneoplastic).

monotherapy, although benefits have been reported in some patients when associated with 3,4-diaminopyridine (3,4-DAP) [29].

In patients in whom the symptoms of the disease interfere with aspects of the day and limitation of daily activities, 3,4-DAP can be used for symptomatic improvement and is recommended, eventually, even as initial therapy. This medication has as its mechanism of action the blockade of potassium channels, which prolongs the depolarization of the presynaptic nerve terminal and, as a consequence, increases the entry of calcium through the VGCC channels, leading to an increase in the release of ACh. The beneficial effects of 3,4-DAP include improvement in muscle strength, autonomic symptoms and CMAP amplitude [29–31]. Sanders et al. [32], in their study carried out in 2018, demonstrated the effectiveness of treatment with medication in symptomatic control.

If the neuromuscular disease is refractory to the two aforementioned medications, immunomodulatory therapy may be an option, with intravenous immunoglobulin being the first line, in addition to other alternatives, such as prednisone, rituximab, azathioprine and plasmapheresis [7].

Intravenous immunoglobulin, therefore, has been successfully used for the treatment of several neurological diseases that present immunomediation, and can be used both as induction treatment in symptomatic patients and for maintenance treatment in patients with recurrence of symptoms [33].

In a double-early, randomized, placebo-controlled crossover study, Bain PG et al. demonstrated significant improvements in muscle, respiratory and bulbar strength indices that were associated with a reduction in serum VGCC antibody patterns [34].

8. Prognosis

In evaluating the survival and prognosis of the patient who has a case of LEMS, the main determinant to be considered is the presence or absence of neoplasia, that is, whether we are facing a paraneoplastic LEMS or not. In patients with SCLC associated with LEMS, a longer survival can be noted, when compared to patients with SCLC, but without the neurological pathology together [35]. In patients without associated neoplasia, survival is normal or almost normal, as seen in the study conducted by Maddison et al. [35], in which 47 patients with LEMS without SCLC, 10 died at a mean age of 70 years, unrelated to the LEMS.

9. Conclusion

Despite being a rare disease in the population context, when referring to the clinical practice of neurology, this pathology is one of the main differential diagnoses in the involvement of the neuromuscular junction. Proximal muscle weakness, autonomic disorders and areflexia are the main clinical findings in the range of diseases that involve this group, which, for the correct diagnosis, must take LEMS into account and be involved in the differential diagnosis.

In case of clinical suspicion, tools that help in the diagnosis are accessible in our midst, in addition to clinical and electrophysiological criteria. Early recognition and immediate initiation of treatment improve patient survival and prognosis, in addition to rapid visualization of the underlying etiology, such as treatment of neoplasms, in case this becomes the cause of the syndrome.

Finally, patients with a suspected or confirmed diagnosis should be covered and examined by a multidisciplinary team, which includes, in addition to neurological treatment, the presence of respiratory, oncological, thoracic, pathological and radiological services. In addition, it is important to highlight the importance of oncological screening and evaluation for the diagnosis of the syndrome, since this disease is closely associated with paraneoplastic involvement and, in some situations, may precede the malignant diagnosis itself.

Acknowledgements

We would like to give due recognition to Dr. Felipe Ibiapina for making this work possible and to the University of Joinville for encouraging and opening doors for such a publication.

Conflict of interests

The authors declare no conflict of interest.

Author details

Felipe Fanine de Souza*, Julia Petry Trevisani and Felipe Ibiapina dos Reis
University of Joinville Region (UNIVILLE), Joinville, SC, Brazil

*Address all correspondence to: ffanine@gmail.com

References

[1] Ivanovski T, Miralles F. Lambert-Eaton myasthenic syndrome: early diagnosis is key. Degener Neurol Neuromuscul Dis. 2019;**9**:27-37. DOI: 10.2147/DNND.S192588. PMID: 31191084; PMCID: PMC6524763

[2] Pascuzzi RM, Bodkin CL. Myasthenia Gravis and Lambert-Eaton myasthenic syndrome: new developments in diagnosis and treatment. Neuropsychiatric Disease and Treatment 2022;**18**:3001-3022. DOI: 10.2147/NDT.S296714. PMID: 36578903; PMCID: PMC9792103

[3] Wirtz PW, Nijnuis MG, Sotodeh M, et al. The epidemiology of myasthenia gravis, Lambert-Eaton myasthenic syndrome and their associated tumors in the northern part of the province of South Holland. Journal of Neurology. 2003;**250**(6):698-701

[4] Sanders DB. Lambert-Eaton myasthenic syndrome: diagnosis and treatment. Ann NY Acad Sci. 2003;**998**:500-508

[5] Wirtz PW, van Dijk JG, van Doorn PA, et al. The epidemiology of Lambert-Eaton myasthenic syndrome in the Netherlands. Neurology. 2004;**63**(2):397-398

[6] Titular MJ, Maddison P, Sont JK, et al. The Lambert-Eaton Dutch-English myasthenic syndrome (LEMS) clinical tumor association score accurately predicts small cell lung cancer in LEMS. Journal of Clinical Oncology. 2011;**29**:902-908

[7] Jayarangaiah A. Lambert eaton myasthenic syndrome. StatPearls. Available from: https://www.ncbi.nlm.nih.gov/books/NBK507891/ [Internet]. Published in January 202

[8] Schoser B, Eymard B, Datt J, Mantegazza R. Lambert-Eaton myasthenic syndrome (LEMS): a rare autoimmune presynaptic disorder often associated with cancer. Journal of Neurology. 2017;**264**(9):1854-1863. DOI: 10.1007/s00415-017-8541-9

[9] Wirtz PW, Smallegange TM, Wintzen AR, et al. Differences in clinical features between Lambert-Eaton myasthenic syndrome with and without cancer: an analysis of 227 published cases. Clinical Neurology and Neurosurgery. 2002;**104**:359-363

[10] Titulaer MJ, Verschuuren JJ. Lambert-Eaton myasthenic syndrome: tumoral versus non-tumour forms. Ann NY Acad Sci. 2008

[11] O'Neill JH, Murray NMF, Newsom-Davis J. Lambert-Eaton myasthenic syndrome: a review of 50 cases. Brain. 1988;**111**:577-596

[12] Meriney SD, Hulsizer SC, Lennon VA, et al. Lambert-Eaton myasthenic syndrome immunoglobulins react with several types of calcium channels in small cell lung carcinoma. Annals of Neurology. 1996;**40**:739-749

[13] Zalewski NL, Lennon VA, Lachance DH, et al. Antibodies to P/Q and N-type calcium channels: oncological, neurological and serological follow-ups. Muscular Nerve. 2016;**54**(2):220-227

[14] Katz E, Ferro PA, Weiss G, et al. Calcium channels involved in synaptic transmission at the neuromuscular junction of mature and regenerating mice. The Journal of Physiology. 1996;**497**:687-689

[15] Titulaer MJ, Lang B, Verschuuren JJ. Lambert-Eaton myasthenic syndrome: from clinical features to therapeutic strategies. Lancet Neurology. 2011;**10**(12):1098-1107

[16] Titulaer MJ, Wirtz PW, Kuks JB, et al. Lambert-Eaton myasthenic syndrome 1988-2008: a clinical picture in 97 patients. Journal of Neuroimmunology. 2008;**201-02**:153-158

[17] Burns TM, Russell JA, LaChance DH, et al. Oculobulbar involvement is typical of Lambert-Eaton myasthenic syndrome. Annals of Neurology. 2003;**53**:270-273

[18] Young JD, Leavitt JA. Lambert Eaton myasthenic syndrome: ocular signs and symptoms. J Neuroftalmol. 2016;**36**:20-22

[19] Cho JJ, Wymer JP. Lambert-Eaton paraneoplastic myasthenic syndrome with limbic encephalitis: clinical correlation with coexistence of anti-VGCC and anti-GABAB receptor antibodies. Journal of Clinical Neuromuscular Disease. 2017;**19**(2):84-88. DOI: 10.1097/CND.0000000000000192

[20] Kitanosono H, Shiraishi H, Motomura M. P/Q-type calcium channel antibodies in Lambert-Eaton myasthenic syndrome. Brain Nerve. 2018;**70**(4):341-355. Japanese. DOI: 10.11477/mf.1416201007. PMID: 29632282

[21] Sun X, Tan J, Sun H, Liu Y, Guan W, Jia J, et al. Anti-SOX1 antibodies in paraneoplastic neurological syndrome. Journal of Clinical Neurology. 2020;**16**(4):530-546. DOI: 10.3988/jcn.2020.16.4.530. PMID: 33029958; PMCID: PMC7541980

[22] Maddison P et al. Detection of neuronal antibodies and better prediction of lung cancer in Lambert-Eaton myasthenic syndrome. Journal of Neuroimmunology. 2020;**340**:577149

[23] Tim RW, Massey JM, Sanders DB. Lambert-Eaton myasthenic syndrome (LEMS): clinical and electrodiagnostic features and response to therapy in 59 patients. Ann NY Acad Sci. 1998;**841**

[24] Oh SJ, Kurokawa K, Claussen GC, Ryan HF. Electrophysiological diagnostic criteria for Lambert-Eaton myasthenic syndrome. Muscular Nerve. 2005;**32**(4):515-520

[25] Hatanaka Y, Oh SJ. Ten-second exercise is superior to 30-second exercise for post-exercise facilitation in diagnosing Lambert-Eaton myasthenic syndrome. Muscle & Nerve. 2008;**37**(5):572-575

[26] Oh SJ, Hatanaka Y, Claussen GC, Sher E. Electrophysiological differences in seropositive and seronegative Lambert-Eaton myasthenic syndrome. Muscle & Nerve. 2007;**35**(2):178-183

[27] Yoon CH, Owusu-Guha J, Smith A, Buschur P. Amifampridine for the treatment of Lambert-Eaton myasthenic syndrome: a new approach to an old drug. Ana Farmacêutica. 2020;**54**(1):56-63

[28] Anwar A, Saleem S, Ahmed MF, Ashraf S, Ashraf S. Recent advances and therapeutic options in Lambert-Eaton myasthenic syndrome. Cureu. 2019;**11**(8)

[29] Wirtz P, Verschuuren J, Dijk JV, et al. Efficacy of 3,4-diaminopyridine and pyridostigmine in the treatment of Lambert-Eaton myasthenic syndrome: a randomized, double-blind, placebo-controlled, crossover study. Clinical Pharmacology and Therapeutics. 2009;**86**(1):44-48

[30] Oh SJ, Claussen GG, Hatanaka Y, Morgan MB. 3,4-Diaminopyridine is more effective than placebo in a randomized, double-blind, cross-over drug study in LEMS. Muscle & Nerve. 2009;**40**(5):795-800

[31] Oh SJ, Shcherbakova N, Kostera-Pruszczyk A, et al. Amifampridine phosphate (Firdapse®) is effective and safe in a phase 3 clinical trial in LEMS. Muscle & Nerve. 2016;**53**(5):717-725

[32] Sanders DB, Juel VC, Harati Y, et al. 3,4-Diaminopyridine base effectively treats the weakness of Lambert-Eaton myasthenia. Muscle & Nerve. 2018;**57**(4):561-568

[33] Rich MM, Teener JW, Bird SJ. Treatment of Lambert-Eaton syndrome with intravenous immunoglobulin. Muscle & Nerve. 1997;**20**(5):614-615

[34] Bain PG, Motomura M, Newsom-Davis J, et al. Effects of intravenous immunoglobulin on muscle weakness and calcium-channel autoantibodies in the Lambert-Eaton myasthenic syndrome. Neurology. 1996;**47**(3):678-683

[35] Maddison P, Gozzard P, Grainge MJ, Lang B. Long-term survival in paraneoplastic Lambert-Eaton myasthenic syndrome. Neurology. 2017;**88**:1334-1339